Problem Books in Mathematics

Edited by P. R. Halmos

Problem Books in Mathematics

Series Editor: P.R. Halmos

Polynomials
by *Edward J. Barbeau*

Problems in Geometry
by *Marcel Berger, Pierre Pansu, Jean-Pic Berry and Xavier Saint-Raymond*

Problem Book for First Year Calculus
by *George W. Bluman*

Exercises in Probability
by *T. Cacoullos*

An Introduction to Hilbert Space and Quantum Logic
by *David W. Cohen*

Unsolved Problems in Geometry
by *Hallard T. Croft, Kenneth J. Falconer, and Richard K. Guy*

Problems in Analysis
by *Bernard Gelbaum*

Theorems and Counterexamples in Mathematics
by *Bernard R. Gelbaum and John M.H. Olmsted*

Exercises in Integration
by *Claude George*

Algebraic Logic
by *S.G. Gindikin*

Unsolved Problems in Number Theory
by *Richard K. Guy*

An Outline of Set Theory
by *James M. Henle*

Demography Through Problems
by *Nathan Keyfitz and John A. Beekman*

(continued after index)

Unsolved Problems in Intuitive Mathematics
Volume II

Hallard T. Croft   Kenneth J. Falconer
Richard K. Guy

Unsolved Problems in Geometry

With 66 Figures

Springer-Verlag
New York   Berlin   Heidelberg   London
Paris   Tokyo   Hong Kong   Barcelona

Hallard T. Croft
Peterhouse,
Cambridge, CB2 1RD
England

Richard K. Guy
Department of Mathematics and Statistics
The University of Calgary
Calgary, Alberta T2N 1N4
Canada

Kenneth J. Falconer
School of Mathematics
Bristol University
Bristol, BS8 1TW
England

Series Editor:
Paul R. Halmos
Department of Mathematics
Santa Clara University
Santa Clara, CA 95053
USA

AMS Classification: 52-02

Library of Congress Cataloging-in-Publication Data
Croft, Hallard T.
    Unsolved problems in geometry / Hallard T. Croft, Kenneth J.
Falconer, Richard K. Guy.
    p. cm. — (Problem books in mathematics) (Unsolved problems
in intuitive mathematics : v. 2)
    Includes bibliographical references and index.
    ISBN 0-387-97506-3 (Springer-Verlag New York Berlin Heidelberg), —
    ISBN 3-540-97506-3 (Springer-Verlag Berlin Heidelberg New York)
    1. Geometry.   I. Falconer, K. J., 1952– .   II. Guy, Richard K., 1916–
III. Title.   IV. Series.   V. Series: Guy, Richard K. Unsolved
problems in intuitive mathematics : v. 2.
QA43.G88   vol. 2
[QA447]
510 s—dc20   90-25041
[516]

Printed on acid-free paper.

Typeset by Asco Trade Typesetting Ltd., Hong Kong.
Printed and bound by R. R. Donnelley & Sons, Harrisonburg, VA.
Printed in the United States of America.

9 8 7 6 5 4 3 2 1

ISBN 0-387-97506-3 Springer-Verlag New York Berlin Heidelberg
ISBN 3-540-97506-3 Springer-Verlag Berlin Heidelberg New York

Preface

For centuries, mathematicians and non-mathematicians alike have been fascinated by geometrical problems, particularly problems that are "intuitive," in the sense of being easy to state, perhaps with the aid of a simple diagram. Sometimes there is an equally simple but ingenious solution; more often any serious attempt requires sophisticated mathematical ideas and techniques. The geometrical problems in this collection are, to the best of our knowledge, unsolved. Some may not be particularly difficult, needing little more than patient calculation. Others may require a clever idea, perhaps relating the problem to another area of mathematics, or invoking an unexpected technique. Some of the problems are very hard indeed, having defeated many well-known mathematicians.

It is hoped that this book will be appreciated at several levels. The research mathematician will find a supply of problems to think about, and should he or she decide to make a serious attempt to solve some of the them, enough references to discover what is already known about the problem. By becoming aware of the state of knowledge on certain topics, mathematicians may notice links with their own special interest, leading to progress on the problem or on their own work. More generally, the book may provide stimulus as "background reading" for mathematicians or others who wish to keep up to date with the state of the art of geometrical problems.

For the interested layman, the book will give an idea of some of the problems that are being tackled by mathematicians today. This may lead to having a go at some of the problems, discovering the difficulties, and perhaps producing a solution, either fallacious or, hopefully, valid. In the area of tiling, for example (see Chapter C), patience and, sometimes, the use of a home computer, have led amateur mathematicians to impressive results.

Perhaps this book will convince others that some of the ideas of mathematics are not wholly beyond their reach, in that the problems, at least, can be understood. This may well lead to the question, "Why are mathematicians wasting their time on such esoteric problems?" One justification might be that, even if the problems themselves have no direct application, the mathematics developed while trying to solve them can be very useful. For example, attempts at proving the (rather useless) Four Color Conjecture led to the development of Graph Theory, a subject of great practical application.

In many developed countries the number of young people embarking on careers of a mathematical nature is dwindling; it is hoped that books such as this may do a little to show them that mathematics *is* a fascinating subject.

Each section in the book describes a problem or a group of related problems. Usually the problems are capable of generalization or variation in many directions; hopefully the adept reader will think of such variations. Frequently, problems are posed in the plane, but could equally well be asked in 3- (or higher-) dimensional space (often resulting in a harder problem). Alternatively, a question about convex sets, say, might be asked about the *special* case of centro-symmetric convex sets or perhaps in the more *general* situation of completely arbitrary sets, resulting in a problem of an entirely different nature.

For convenience, the problems have been divided into seven chapters. However, this division and the arrangement within the chapters is to some extent arbitrary. The many interrelationships between the problems make a completely natural ordering impossible.

References for each group of problems are collected at the end of the section. A few, indicated by square brackets [ ], refer to the list of Standard References at the front of the book; this is done to avoid excessive repetition of certain works. Where useful, we have included the number of the review in Mathematical Reviews, prefixed by MR. Some sections have few references, others a large number. Not all problems have a complete bibliography—a full list of articles that relate to Sylvester's Problem (F12), for example, would fill the book. However, we have done our best to include those books and papers likely to be most helpful to anyone wishing to find out more about a problem, including the most important and recent references and survey articles. Obviously, these will, in turn, lead to further references. Some books and papers listed are not referred to directly in the text but are nonetheless relevant. There are inevitably some omissions, for which we apologize.

This book has a long history, and many correspondents will have despaired of its ever appearing. Early drafts date back to about 1960 when HTC started to collect problems that were unsolved but easy to state, particularly problems on geometry, number theory, and analysis. As a frequent visitor to Cambridge, RKG became aware of the collection and started contributing to it. Interest was reinforced at the East Lansing Geometry Conference in March 1966, in which HTC, RKG, John Conway, Paul Erdős, and Leo Moser participated, and where Moser circulated his "Fifty poorly posed problems in combinatorial geometry." Encouraged by HTC, Erdős, and Moser, RKG started to formalize these lists into a book. The mass of material became far too large for a single volume, with the Number Theory "chapter" resembling a book in its own right. This was published in 1981 as "Unsolved Problems in Number Theory," Volume I of this series.

The geometry and analysis drafts were regularly updated by HTC until about 1970, when the mass of material collected was enormous. However, this part of the project became dormant until KJF, a research student of HTC, became involved in 1986. By this time, progress had been made on the solution of many of the problems, and it was left to KJF with some help from RKG to update and rewrite completely most of the sections and to add further problems to the collection.

There are still many problems that we have collected which would be nice to share. We have in mind (should we live long enough) Volume III on Combinatorics, Graphs, and Games and Volume IV on Analysis.

We are most grateful to the many people who have, over the past 30 years, corresponded and sent us problems, looked at parts of drafts, and made helpful comments. These include G. Blind, D. Chakerian, John Conway, H. S. M. Coxeter, Roy Davies, P. Erdős, L. Fejes Tóth, Z. Füredi, R. J. Gardner, Martin Gardner, Ron Graham, B. Grünbaum, L. M. Kelly, V. Klee, I. Leader, L. Mirsky, Leo Moser, Willy Moser, C. M. Petty, R. Rado, C. A. Rogers, L. A. Santaló, Jonathan Schaer, John Selfridge, H. Steinhaus, G. Wengerodt, and J. Zaks. The bulk of the technical typing was done by Tara Cox and Louise Guy. The figures were drawn by KJF using a combination of a computer graphics package and freehand. The staff at Springer-Verlag in New York have been courteous, competent, and helpful.

In spite of this help, many errors and omissions remain. Some of the problems may have solutions that are unpublished or in papers that we have overlooked, others will doubtless have been solved since going to press. There must be many references that we are unaware of. The history of some of the problems may have become forgotten or confused—it is often unclear who first thought of a particular problem, and many problems undoubtedly occur to several people at about the same time. This book will no doubt perpetuate such confusion. We can do no more than offer our reluctant apologies for all this. We would be glad to hear of any omissions or corrections from readers, so that any future revision may be more accurate. Please send such comments to KJF.

Cambridge, Bristol and Calgary, September 1990   HTC, KJF, RKG

Other Problem Collections

Here are some good sources of general geometrical problems, many of which are lists compiled at conference meetings. Collections of problems on more specific subjects are mentioned in the chapter introductions.

P. Erdős, Some combinatorial problems in geometry, Lecture Notes in Mathematics, 792, Springer-Verlag, New York, 1980, 46–53; *MR* **82d**:51002.

W. Fenchel (ed.), Problems, [Fen] 308–325.

P. M. Gruber & R. Schneider, *Problems in Geometric Convexity*, [TW] 255–278.

R. K. Guy (ed.), Problems, in *The Geometry of Metric and Linear Spaces*, Lecture Notes in Mathematics 490, Springer-Verlag, New York, 1974, 233–244.

D. C. Kay & M. Breen (eds.), [KB] 229–234.

V. L. Klee (ed.), Unsolved problems, [Kle] 495–500.

V. L. Klee, Some unsolved problems in plane geometry, *Math. Mag.* **52** (1979) 131–145; *MR* **80m**:52006.

D. G. Larman & C. A. Rogers (eds.), Problems, Durham Symposium on relations between finite and infinite dimensional convexity, *Bull. London Math. Soc.* **8** (1976) 28–33.

J. E. Littlewood, *Some Problems in Real and Complex Analysis*, Heath, Boston, 1968; *MR* **39** #5777.

R. D. Mauldin (ed.), *The Scottish Book*, [Mau].

Z. A. Melzak, Problems conerned with convexity, *Canad. Math. Bull.* **8** (1965) 565–573; *MR* **33** #4781.

Z. A. Melzak, More problems concerned with convexity, *Canad. Math. Bull.* **11** (1968) 482–494; *MR* **38** #3767.

H. Meschkowski, *Unsolved and Unsolvable Problems in Geometry*, [Mes].

W. Moser & J. Pach, *Research Problems in Discrete Geometry*, [MP].

C. S. Ogilvy, *Tomorow's Math*, Oxford University Press, Oxford, 1962.

H. Steinhaus, *One Hundred Problems in Elementary Mathematics*, Pergamon, Oxford, 1964; *MR* **28** #1110.

S. Ulam, *A Collection of Mathematical Problems*, Interscience, New York, 1960; *MR* **22** #10884.

Journals that publish problems on a regular basis include: *American Mathematical Monthly, Colloquium Mathematicum, Elemente der Mathematik, Mathematical Intelligencer, Mathematics Magazine, SIAM Review*, and *Crux Mathematicarum*.

Standard References

The following books and conference proceedings are referred to so frequently that we list them here, and refer to them by letters in square brackets throughout the book.

[Bla] W. Blaschke, *Kreis und Kugel*, 1916; Chelsea reprint, New York 1949; *MR* **17**, 887.

[BonFe] T. Bonnesen & W. Fenchel, *Theorie der Konvexer Körper*, Springer, Berlin, 1934; Chelsea reprint, New York 1971; *MR* **51** #8954.

[BF] K. Böröczky & G. Fejes Tóth (eds.), *Intuitive Geometry, Siófok, 1985*, North-Holland, Amsterdam, 1987; *MR* **88h**:52002.

[CS] J. H. Conway & N. J. A. Sloane, *Sphere Packings, Lattices and Groups*, Springer-Verlag, New York, 1988.

[DGK] L. Danzer, B. Grünbaum & V. Klee, *Helly's Theorem and its Relatives*, in *Convexity* (V. L. Klee, ed.), Proc. Symp. Pure Math. 7, Amer. Math. Soc., Providence, 1963, 101–180; *MR* **28** #524.

[DGS] C. Davis, B. Grünbaum & F. A. Sherk (eds.), *The Geometric Vein—The Coxeter Festschrift*, Springer-Verlag, New York, 1981; *MR* **83e**:51003.

[Fej] L. Fejes Tóth, *Lagerungen in der Ebene, auf der Kugel und in Raum*, 2nd ed., Springer, Berlin, 1972; *MR* **50** #5603.

[Fej'] L. Fejes Tóth, *Regular Figures*, Pergamon, Oxford, 1954.

[Fen] W. Fenchel (ed.), *Proceedings of the Colloquium on Convexity, 1965*, Københavns Univ. Mat. Inst., 1967; *MR* **35** #5271.

[GLMP] J. E. Goodman, E. Lutwak, J. Malkevitch & R. Pollack (eds.), Discrete geometry and convexity, *Ann. New York Acad. Sci.* **440** (1985); *MR* **86g**:52002.

[Gru] B. Grünbaum, *Convex Polytopes*, Interscience, New York and London, 1967; *MR* **37** #2085.

[GS] B. Grünbaum & G. C. Shephard, *Tilings and Patterns*, Freeman, New York, 1987; *MR* **88k**:52018. (The first seven Chapters are available separately as "*Tilings and Patterns—An Introduction*.")

[GW] P. M. Gruber & J. M. Wills (eds.), *Convexity and its Applications*, Wien 1981, Siegen, 1982, Birkhäuser, Basel, 1983; *MR* **84m**:52001.

[Had] H. Hadwiger, *Altes und Neues über Konvexer Körper*, Birkhäuser, Basel, 1955; *MR* **17**, 401.

[Ham] J. Hammer, *Unsolved Problems Concerning Lattice Points*, Pitman, London, 1977; *MR* **56** #16515.

[HDK] H. Hadwiger & H. Debrunner (translated by V. Klee), *Combinatorial Geometry in the Plane*, Holt, Rinehart, and Winston, New York, 1964; *MR* **29** #1577.

[Kay] D. C. Kay (ed.), Proc. Conf. on Convexity and Combinatorial Geometry, Norman, Oklahoma, 1971, University of Oklahoma, 1971; *MR* **48** #9547.

[KB] D. C. Kay & M. Breen (eds.), *Conference on Convexity and Related Combinatorial Geometry, Oklahoma, 1980*, Marcel Dekker, New York, 1982; *MR* **83b** #52002.

[Kla] D. A. Klarner (ed.), *The Mathematical Gardner*, Wadsworth, Belmont, 1981; *MR* **82b**:00003.

[Kle] V. L. Klee (ed.), Convexity, Proc. Symp. Pure Math. 7, Amer. Math. Soc., Providence, 1963.

[Mau] R. D. Mauldin (ed.), *The Scottish Book*, Birkhäuser, Basel, 1981; *MR* **84m**:00015.

[Mes] H. Meschkowski, *Unsolved and Unsolvable Problems in Geometry*, Oliver and Boyd, 1966; *MR* **35** #7206.

[MP] W. Moser & J. Pach, Research problems in discrete geometry, Mineographed Notes, 1985.

[Rog] C. A. Rogers, *Packing and Covering*, Cambridge University Press, Cambridge, 1964; *MR* **30** #2405.

[RZ] M. Rosenfeld & J. Zaks (eds.), *Proceedings of Jerusalem Conference, 1981*, Ann. Discrete Math. 20, North-Holland, Amsterdam, 1984; *MR* **86g**:52001.

[TW] J. Tölke & J. M. Wills (eds.), *Contributions to Geometry*, Birkhäuser, Basel, 1979; *MR* **81b**:52002.

Contents

B. Polygons, Polyhedra, and Polytopes   48

C. Tiling and Dissection   79

D. Packing and Covering   107

E. Combinatorial Geometry   131

F. Finite Sets of Points   149

G. General Geometric Problems   168

Notation and Definitions

The following brief synopsis serves to introduce some of the notation and terminology that will be needed throughout the book. More specific notions are discussed in the chapter introductions, or, in some cases, in the individual sections.

Sets

We shall require some definitions and notation from set theory.

Most of our problems are posed in d-**dimensional Euclidean space**, $\mathbb{R}^d$; in particular $\mathbb{R}^1 = \mathbb{R}$ is just the set of real numbers or the real line, $\mathbb{R}^2$ is the (Euclidean) plane, and $\mathbb{R}^3$ is usual (Euclidean) space. Points in $\mathbb{R}^d$ are printed in bold type $\mathbf{x}$, $\mathbf{y}$, etc, and we will sometimes use the coordinate form $\mathbf{x} = (x_1, \ldots, x_d)$. If $\mathbf{x}$ and $\mathbf{y}$ are points of $\mathbb{R}^d$, the **distance** between them is $|\mathbf{x} - \mathbf{y}| = (\sum_{i=1}^d |x_i - y_i|^2)^{1/2}$.

Sets, which will generally be subsets of $\mathbb{R}^d$, are denoted by capital letters (e.g., E, F, K, etc). In the usual way, $\mathbf{x} \in E$ means that the point $\mathbf{x}$ is a member of the set E, and $E \subset F$ means that E is a subset of F. We write $\{\mathbf{x}: \text{condition}\}$ for the set of $\mathbf{x}$ for which "condition" is true. The **empty set**, which contains no elements, is written $\varnothing$. The set of integers is denoted by $\mathbb{Z}$ and the rational numbers by $\mathbb{Q}$. We sometimes use a superscript $^+$ to denote the positive elements of a set (e.g., $\mathbb{R}^+$ is the set of positive real numbers).

The **closed ball** of center $\mathbf{x}$ and radius r is defined by $B_r(\mathbf{x}) = \{\mathbf{y}: |\mathbf{y} - \mathbf{x}| \le r\}$. Similarly, the **open ball** is $\{\mathbf{y}: |\mathbf{y} - \mathbf{x}| < r\}$. Thus the closed ball contains its bounding sphere, but the open ball does not. Of course, in $\mathbb{R}^2$ a ball is a disk, and in $\mathbb{R}^1$ a ball is just an interval. If $a < b$, we write $[a, b]$ for the **closed interval** $\{x: a \le x \le b\}$ and (a, b) for the **open interval** $\{x: a < x < b\}$.

We write $E \cup F$ for the **union** of the sets E and F (i.e., the set of points belonging to either E or F). Similarly, we write $E \cap F$ for their **intersection** (i.e., the points in both E and F). More generally $\bigcup_i E_i$ denotes the **union** of an arbitrary collection of sets $\{E_i\}$ (i.e., those points in at least one E_i) and $\bigcap_i E_i$ denotes their **intersection**, consisting of the points common to all of the sets E_i. A collection of sets is **disjoint** if the intersection of any pair is the empty set. The **difference** $E \backslash F$ consists of those points in E that are not in F, and $\mathbb{R}^d \backslash E$ is called the **complement** of E.

An infinite set E is **countable** if its elements can be listed in the form $x_1, x_2, \ldots$ with every element of E appearing at a specific place in the list; otherwise the set is **uncountable**. The sets $\mathbb{Z}$ and $\mathbb{Q}$ are countable but $\mathbb{R}$ is uncountable.

If E is any set of real numbers, the **supremum**, sup E, is the least number m such that $x \leq m$ for every x in E. Similarly, the **infimum**, inf E, is the greatest number m such that $m \leq x$ for every x in E. Roughly speaking, we think of inf E and sup E as the minimum and maximum of the numbers in E, though it should be emphasized that inf E and sup E need not themselves be in E.

We use the "floor" and "ceiling" symbols "$\lfloor \ \rfloor$" and "$\lceil \ \rceil$" to mean "the greatest integer not more than" and "the least integer not less than."

The **diameter**, diam E, of a subset E of $\mathbb{R}^d$ is the greatest distance apart of pairs of points in E; thus diam $E = \sup\{|x - y| : x, y \in E\}$. A set A is **bounded** if it has finite diameter, or, equivalently, is contained in some (sufficiently large) ball.

We have already used the terms "open" and "closed" in connection with intervals and balls, but these notions extend to much more general sets. Intuitively, a set is closed if it contains its boundary and open if it contains none of its boundary points. More precisely, a subset E of $\mathbb{R}^d$ is **open** if, for every x in E, there is some ball $B_r(x)$ of positive radius r, centered at x and contained in E. A set E is **closed** if its complement is open; equivalently if for every sequence x_r in E that is convergent to a point x of $\mathbb{R}^d$, we have $x \in E$. The empty set $\varnothing$ and $\mathbb{R}^d$ are regarded as both open and closed. The union of any collection of open sets is open, as is the intersection of a *finite* collection of open sets. The intersection of any collection of closed sets is closed, as is the union of a *finite* number of closed sets.

The smallest closed set containing a set E, more precisely, the intersection of all closed sets that contain E, is called the **closure** of E. Similarly, the **interior** of a set E is the largest open set contained in E, that is the union of all open subsets of E. The **boundary** of E is defined as the set of points in the closure of E but not in its interior.

For our purposes, a subset of $\mathbb{R}^d$ is **compact** if it is closed and bounded.

A set E is thought of as connected if it consists of just one "piece"; formally E is **connected** if there do not exist open sets U and V such that $U \cup V$ contains E and with $E \cap U$ and $E \cap V$ disjoint and nonempty. A subset E of $\mathbb{R}^2$ is termed **simply connected** if both E and $\mathbb{R}^2 \backslash E$ are connected.

There is a further class of sets that will be mentioned occasionally, though its precise definition is indirect, and need not unduly concern the reader. The

Borel sets are, roughly speaking, the sets that can be built up from open or closed sets by repeatedly taking countable unions and intersections. More precisely, the class $\mathscr{B}$ of **Borel sets** in $\mathbb{R}^d$ is the smallest collection of sets that includes the open and closed sets, such that if $E, E_1, E_2, \ldots$ are in $\mathscr{B}$ then so are $\bigcup_{i=1}^{\infty} E_i$, $\bigcap_{i=1}^{\infty} E_i$ and $\mathbb{R}^d \backslash E$.

Occasionally we need to indicate the degree of smoothness of a curve or surface. We say that such a set is C^k ($k = 1, 2 \ldots$) if it may be defined locally, with respect to suitable coordinate axes, by a function that is k times differentiable with continuous kth derivative. A curve or surface is C^{∞} if it is C^k for every positive integer k.

The notation $f(x) = o(g(x))$ means that $f(x)/g(x) \to 0$ as $x \to \infty$, and $f(x) = O(g(x))$ means that there is a constant c such that $|f(x)| \leq c|g(x)|$ for all sufficiently large x. Similarly, $f(x) \sim g(x)$ means that $f(x)/g(x) \to 1$ as $x \to \infty$.

Geometrical transformations

Let E and F be any sets. A **mapping, function,** or **transformation** f from E to F is a rule or formula that associates a point $f(x)$ of F with each point x of E. We write $f: E \to F$ to denote this situation. If $A \subset E$, we write $f(A) = \{f(x): x \in A\}$ for the **image** of A.

A function $f: E \to F$ is called an **injection** or **one-to-one** function if $f(x) \neq f(y)$ whenever $x \neq y$ (i.e., if different elements of E are mapped to different elements of F). A function is called a **surjection** or an **onto** function if, for every $y \in F$, there exists $x \in E$ such that $f(x) = y$. A function that is both an injection and a surjection is called a **bijection** or a **one-to-one correspondence** between E and F.

Certain transformations have particular geometric significance. A transformation $S: \mathbb{R}^d \to \mathbb{R}^d$ is called a **congruence** or **isometry** if it preserves distances (i.e., if $|S(x) - S(y)| = |x - y|$ for all $x, y \in \mathbb{R}^d$). Such a transformation also preserves angles and transforms sets into **congruent** ones. Special cases include **translations,** which shift points a constant distance in parallel directions, **rotations,** which have a center a such that $|S(x) - a| = |x - a|$ for all x, and **reflections,** which map all points to their mirror images in a fixed $(d - 1)$-dimensional plane. A congruence that may be achieved by a translation followed by a rotation is sometimes called a **rigid motion** or **direct congruence**. A transformation S is a **similarity** if there is a positive constant c such that $|S(x) - S(y)| = c|x - y|$ for all $x, y \in \mathbb{R}^d$, and transforms each set E into a **similar** set $S(E)$. A similarity that preserves orientation (i.e., for which the line segments $[S(x), S(y)]$ and $[x, y]$ are parallel) is called a **homothety** and E and $S(E)$ are termed **homothetic**. An **affinity** or **affine transformation** transforms straight lines to straight lines and may be thought of as a shearing transformation; the contracting or expanding effect need not be the same in every direction. The effect of these transformations is shown in Figure N1.

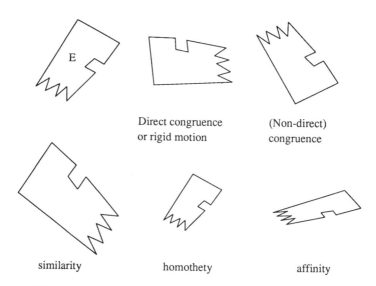

Direct congruence   (Non-direct)
or rigid motion   congruence

similarity   homothety   affinity

Figure N1.   The effect of various transformations on a set E.

Length, area, and volume

For most of this book, an intuitive idea of length, area, and volume will be perfectly adequate. However, a few problems involve sets that may be rather irregular, and precise formulation of length, area, and volume requires a few ideas from measure theory.

If E is a subset of $\mathbb{R}$, we define the **length** or (one-dimensional) **Lebesgue measure** $L(E)$ of E as the infimum (i.e., the smallest possible value) of the sums $\sum_{i=1}^{\infty} (b_i - a_i)$ over all countable collections of intervals $\bigcup_{i=1}^{\infty} [a_i, b_i]$ that cover E. If E itself consists of a finite or countable collection of intervals, then $L(E)$ equals the sum of the interval lengths. It turns out that it is not possible to define $L(E)$ consistently on all subsets of $\mathbb{R}$, but only on a rather large class of subsets called the **Lebesgue measurable sets**. Intervals, open and closed sets, and Borel sets are all Lebesgue measurable, and finite and countable unions and complements of measurable sets are always measurable. Length is additive, indeed countably additive, on such sets, in the sense that if $E_1, E_2, \ldots$ are disjoint measurable sets, then $L(\bigcup_{i=1}^{\infty} E_i) = \sum_{i=1}^{\infty} L(E_i)$. Any sets that can be constructed "effectively" (i.e., by specifying exactly which points are in the set and not resorting to an axiom such as the axiom of choice) are measurable, so for intuitive purposes, we may think of Lebesgue measure as length in the obvious way.

Similarly, we can make the ideas of area or volume precise by introducing 2- or 3-dimensional Lebesgue measure. Thus if $E \subset \mathbb{R}^2$, we define the **area** or **plane Lebesgue measure** of E to be $A(E)$, the infimum of the sums $\sum_{i=1}^{\infty} (b_i - a_i)(d_i - c_i)$ taken over all countable unions of rectangles

$\bigcup_{i=1}^{\infty} [a_i, b_i] \times [c_i, d_i]$ that cover E. As in the one-dimensional case, $A(E)$ is defined consistently on the familiar types of sets. Volume of sets in $\mathbb{R}^3$, and more generally, d-dimensional volume of subsets of $\mathbb{R}^d$, are defined analogously.

For a full treatment of measure theory, see, for example, Kingman & Taylor.

J. F. C. Kingman & S. J. Taylor, *Introduction to Measure and Probability*, Cambridge University Press, Cambridge, 1966; *MR* **36** #1601.

A. Convexity

The division of problems between this chapter and the others is fairly arbitrary. We have tried to select those problems where convexity is an essential feature. Many of the problems can be posed without requiring convexity, often resulting in a problem of a totally different character. Many of the items in the chapter on Polygons and Polytopes are also "convexity" problems, but restricted to that very special class of convex set.

A set K in d-dimensional space is called a **convex set** or **convex body** if the line segment joining any pair of points of K lies entirely in K, (see Figure A1). Thus, convex sets include circular disks, ellipses, squares, rectangles, spheres, parallelepipeds, and the Platonic solids. While the class of convex sets is very large, convexity is a strong enough condition to provide many interesting properties and problems.

Virtually all convex sets that we encounter will be compact (i.e., bounded and with the boundary regarded as part of the set). Generally K will denote a convex set, with C its bounding curve or surface.

There are a number of basic "measures" associated with convex sets which we list below for reference (see Figure A2).

For plane (2-dimensional) convex sets:

A—**area** (plane Lebesgue measure) of K
L—**perimeter length**.

For convex sets in three or more dimensions:

V—**volume** or d-**dimensional content** (i.e., d-dimensional Lebesgue measure)
S—**surface area** or $(d-1)$-**dimensional surface area** (Formally, this may be defined as $\lim_{\varepsilon \to 0}(V(K_\varepsilon) - V(K))/\varepsilon$, where K_ε is the set of points within distance ε of K, and V is d-dimensional content).

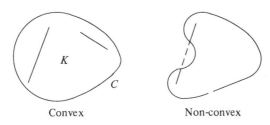

Figure A1.   Convex and non-convex sets.

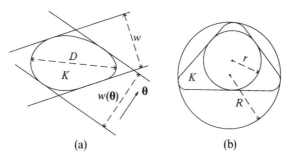

Figure A2.   Measures of a convex set K: (a) diameter D, width $w(\theta)$ in direction θ, and width w (b) inradius r and circumradius R.

For any convex sets:

  D—**diameter** (i.e., the maximum of $|\mathbf{x} - \mathbf{y}|$ as $\mathbf{x}$, $\mathbf{y}$ range through K)

  $w(\theta)$—**width** or **breadth** perpendicular to the unit vector θ, (i.e., the separation of a pair of parallel lines or planes perpendicular to θ touching K on opposite sides)

  w—**width** or **breadth** (i.e., the minimum of $w(\theta)$ over all θ)

  r—**inradius** (i.e., the radius of the largest disk or sphere contained in K the center of which is called the **incenter**)

  R—**circumradius** (i.e., the radius of the smallest disk or sphere containing K, the center of which is called the **circumcenter**)

For convex sets in three dimensions:

  M—the **integral mean curvature** of the surface of K. For smooth surfaces $M = \int \frac{1}{2}(K_1 + K_2)\, dS$, where K_1 and K_2 are the principal curvatures at a surface point and integration is over the surface of K. It may be shown that $M = \frac{1}{2}\int w(\theta)\, d\theta$; this is a rather more convenient form in that it is defined for convex sets that are not necessarily smooth.

  A convex set K is **centro-symmetric** or **centrally symmetric** if it has a **center** $\mathbf{p}$ that bisects every chord of K through $\mathbf{p}$ [see Figure A3(a)]

  A convex set is of **constant width** w if its width is the same in all directions (i.e., $w(\theta) = w$ for all θ). An equivalent condition is $w = D$. The **Reuleaux**

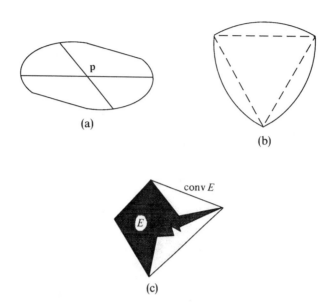

Figure A3.   (a) A centro-symmetric convex set with center **p**. (b) The Reuleux triangle, with each arc centered at the opposite vertex, is a set of constant width. (c) A set E and its convex hull conv E.

triangle, consisting of three circular arcs centered at the opposite vertices [Figure A3(b)] is of constant width.

The **convex hull**, conv E, of any set E is the smallest convex set that contains E [i.e., the intersection of all convex sets containing E, which is necessarily convex, see Figure A.3(c)].

The **dimension**, dim K, of a convex set K in $\mathbb{R}^d$ is the least integer s such that K is contained in an s-dimensional flat (i.e., translate of an s-dimensional subspace). Thus a convex set in 3-dimensional space is 0-dimensional if it is a single point, 1-dimensional if it is a line segment, 2-dimensional if it is contained in a plane but not in a line, and 3-dimensional if it contains a ball of positive radius. It is sometimes convenient to define the dimension of an arbitrary set E as the dimension of its convex hull, and write dim E = dim conv E. Very often questions in convexity are only of interest for **proper** convex sets in $\mathbb{R}^d$, i.e., those with dim $E = d$, or, equivalently, those with nonempty interior.

We list below the standard references on convexity.

R. V. Benson, *Euclidean Geometry and Convexity*, McGraw-Hill, New York, 1966; *MR* **35** #844.

W. Blaschke, [Bla].

T. Bonnesen & W. Fenchel, [BF].

H. Busemann, *Convex Surfaces*, Interscience, New York, 1958; *MR* **21** #3900.

H. G. Eggleston, *Convexity*, Cambridge University Press, Cambridge, 1958; *MR* **23** #A2123.

H. G. Eggleston, *Problems in Euclidean Space—Application of Convexity*, Pergamon, New York, 1957; *MR* **23** #A3228.

L. Fejes Tóth, [Fej].

L. Fejes Tóth, [Fej'].

H. Guggenheimer, *Applicable Geometry—Global and Local Convexity*, Krieger, New York, 1977; *MR* **56** #1198.

H. Hadwiger, [Had].

H. Hadwiger, H. Debrunner & V. Klee, [HDK].

H. Hadwiger, *Vorlesungen über Inhalt, Oberfläche und Isoperimetrie*, Springer, Berlin, 1957; *MR* **21** #1561.

P. J. Kelly & M. L. Weiss, *Geometry and Convexity*, Wiley, New York, 1979; *MR* **80h**:52001.

S. R. Lay, *Convex Sets and their Applications*, Wiley, New York, 1982; *MR* **83e**:52001.

K. Leichtweiss, *Konvexe Mengen*, Springer, Berlin, 1980; *MR* **81j**:52001.

L. A. Lyusternik, *Convex Figures and Polyhedra*, Dover, New York, 1963, Heath, Boston, 1966; *MR* **19**, 57; **28** #4427; **36** #4435.

F. A. Valentine, *Convex Sets*, McGraw-Hill, New York, 1964; *MR* **30** #503.

I. M. Yaglom & V. G. Boltyanskii, *Convex Figures*, Moscow, 1951; English transl. Holt, Rinehart, and Winston, New York, 1961; *MR* **23** #A1283.

The following general articles also provide an introduction to aspects of convexity:

V. Klee, What is a convex set? *Amer. Math. Monthly* **78** (1971) 616–631; *MR* **44** #3202.

J. Dubois, Sur la convexité et ses applications, *Ann. Sci. Math. Québec* **1** (1977) 7–31; *MR* **58** #24002.

P. M. Gruber, Seven small pearls from convexity, *Math. Intelligencer* **5** (1983) No. 1, 16–19; *MR* **85h**:52001.

P. M. Gruber, Aspects of convexity and its applications, *Exposition. Math.* **2** (1984) 47–83; *MR* **86f**:52001.

The following volumes of conference proceedings, which are detailed in full under "Standard References" on pages xi–xii, will be invaluable to any serious student of convexity: [BF], [DGS], [Fen], [GLMP], [GW], [Kay], [Kle], [KB], [RZ], [TW].

A1. The equichordal point problem.

Perhaps the most notorious of all problems in plane convexity was posed by Fujiwara and by Blaschke, Rothe & Weitzenböck in 1917. Is there a plane convex set having two distinct equichordal points? An **equichordal point** has the property that every chord through it has the same length, which we may take to equal 1 (see Figure A4). A number of incorrect "proofs" of the conjecture have been published, but a complete solution still seems far way. To quote Rogers: "If you are interested in studying the problem, my first advice is 'Don't'. My second is 'If you must, do study the work of Wirsing and Butler,' and the third is 'You may well have to develop a sophisticated technique for obtaining uniform and extremely accurate asymptotic expansions for the solutions of a certain recurrence relation giving sequences of points on the boundary of such sets.'"

Quite a bit is known about sets with two equichordal points *if* they exist. Wirsing showed that they must be symmetric about the line L through the two points and also about their perpendicular bisector. He also showed that

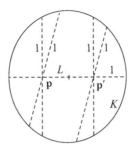

Figure A4.   **p** and **p′** are equichordal points of K, with all chords shown of length one.

the boundary must be a real-analytic curve, and gave a recurrence relationship for the coefficients of its power-series expansion near its intersections with L. It follows that there are at most countably many such sets to within congruence. Calling a the distance between the equichordal points, Ehrhart has shown that $a < 0.5$ and Michelacci has shown that a must be one of a discrete set of numbers with $a < 0.33$. Extensive computation is required in this work.

Wirsing has suggested the following generalized problem: Let C be a closed convex curve, symmetrical about the origin **o**. Let C_+, C_- be C shifted a distance to the right and left, respectively. Let $f_+(\theta)$ and $f_-(\theta)$ be the defining functions for C_+ and C_- in polar coordinates with respect to **o**. Characterize the curves that we can obtain as $r(\theta) = \frac{1}{2}\{f_+(\theta) + f_-(\theta)\}$. If we can get a circle, then C has two equichordal points.

The corresponding problem on the surface of a sphere turns out to be easier. Spaltenstein specified a family of convex sets on the sphere each with two equichordal points. (Here "convexity" and "chord" are defined in terms of segments of great circle arcs). Similarly, Petty & Crotty showed that there are (real) normed spaces in which there are convex sets with two equichordal points.

Klee asked the related question whether there exist nonelliptical convex curves C with two equireciprocal points. (A point **p** is **equireciprocal** if every chord [**x**, **y**] of C through **p** satisfies $|\mathbf{x} - \mathbf{p}|^{-1} + |\mathbf{y} - \mathbf{p}|^{-1} = c$ for some constant c.) Falconer, see also Hallstrom, showed that, except for certain unlikely possibilities, any curve with two equireciprocal points must have the same value of c at each point. Further, any twice differentiable convex curve with two equireciprocal points must be an ellipse, but on the other hand there are nonelliptical convex curves with two equireciprocal points. One can generalize the problem to seek curves with pairs of points satisfying $|\mathbf{x} - \mathbf{p}|^{\alpha} + |\mathbf{y} - \mathbf{p}|^{\alpha} = c$ for any α. Do such curves exist for α other than -1 and 0?

W. Blaschke, W. Rothe & R. Weitzenböck, Aufgabe 552, *Arch. Math. Phys.* **27** (1917) 82.

G. J. Butler, On The "equichordal curve" problem and a problem of packing and covering, PhD thesis, London, 1969.

G. A. Dirac, Ovals with equichordal points, *J. London Math. Soc.* **27** (1952) 429–437; **28** (1953) 256; *M R* **14**, 309.

L. Dulmage, Tangents to ovals with two equichordal points, *Trans. Roy. Soc. Canada Sect. III* (3) **48** (1954) 7–10; *MR* **16**, 740.

E. Ehrhart, Un ovale à deux points isocordes? *Enseign. Math.* **13** (1967) 119–124; *MR* **37** #823.

K. J. Falconer, On the equireciprocal point problem, *Geom. Dedicata* **14** (1983) 113–126; *MR* **84i**:52004.

M. Fujiwara, Über die Mittelkurve zweier geschlossenen konvexen Kurven in Bezug auf einen Punkt, *Tôhoku Math. J.* **10** (1916) 99–103.

R. J. Gardner, Chord functions of convex bodies, *J. London Math. Soc.* (2) **36** (1987) 314–326; *MR* **88h**:52006.

H. Hadwiger, Ungelöste Problem 3, *Elem. Math.* **10** (1955) 10–19.

A. P. Hallstrom, Equichordal and equireciprocal points, *Bogasici Univ. J. Sci.* **2** (1974) 83–88.

V. Klee, Can a plane convex body have two equireciprocal points? *Amer. Math. Monthly* **76** (1969) 54–55, correction **78** (1971) 1114.

D. G. Larman & N. K. Tamvakis, A characterization of centrally symmetric convex bodies in E^n, *Geom. Dedicata* **10** (1981) 161–176; *MR* **82j**:52008.

G. Michelacci, A negative answer to the equichordal problem for not too small excentricities, *Bol. Un. Mat. Ital.* A(7) **2** (1988) 203–211; *MR* **89f**:52003.

C. M. Petty & J. M. Crotty, Characterization of spherical neighbourhoods, *Canad. J. Math.* **22** (1970) 431–435; *MR* **41** #2538.

C. A. Rogers, Some problems in the geometry of convex bodies, in [DGS], 279–284; *MR* **84b**:52001.

C. A. Rogers, An equichordal problem, *Geom. Dedicata* **10** (1981) 73–78; *MR* **82j**:52009.

N. Spaltenstein, A family of curves with two equichordal points on a sphere, *Amer. Math. Monthly* **91** (1984) 423.

W. Suss, Einbereiche mit ausgezeichneten Punkten, Inhalts, und Umfangspunkt, *Tôhoku Math. J.* (1) **25** (1925) 86–98.

E. Wirsing, Zur Analytisität von Doppelspeichkurven, *Arch. Math.* **9** (1958) 300–307; *MR* **21** #2205.

A2. Hammer's x-ray problems.

Suppose a homogeneous solid contains a convex hole K and x-ray photographs are taken so that the "darkness" at each point on a photograph determines the length of the chord of K along an x-ray line [see Figure A5(a)]. How many pictures must be taken to permit exact reconstruction of K if

(a) the x-rays issue from a point source, or
(b) the x-rays are assumed parallel?

In the plane case the problems may be expressed mathematically as follows:

(a′) We are given points $x_1, \ldots, x_k$ and functions $f_1, \ldots, f_k$: $[0, \pi) \to \mathbb{R}$, and seek a compact convex set K such that K intersects the line through x_i making an angle θ with some fixed axis in a chord of length $f_i(\theta)$ [Figure A5(b)]. (We say that K has **chord function** f_i at x_i.) For the sake of generality we allow the x_i to be either interior or exterior to K.

(b′) We are given angles $\theta_1, \ldots, \theta_k$ and functions $F_1, \ldots, F_k$: $\mathbb{R} \to \mathbb{R}$, and seek a compact convex set K such that K intersects the line in the direction θ_i and at perpendicular distance t from the origin in a chord of length $F_i(t)$ [Figure A5(c)].

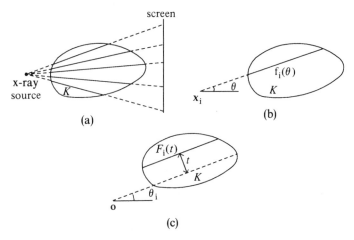

Figure A5. Hammer's x-ray problem. (a) The intensity of the shadow on the screen determines the lengths of chords of K. (b) The point source problem. (c) The parallel chord problem.

There are a number of interesting questions which we state for case (a'); the analogs for (b') should be clear.

(i) Uniqueness. When does a set of points $x_1, x_2, \ldots, x_k$ have the property that at most one convex set K corresponds to any set of chord functions $f_1, \ldots, f_k$?

(ii) Reconstruction. Given that a set of points and chord functions correspond to a unique convex K, reconstruct K from this information.

(iii) Relative uniqueness. Given a convex set K, find a "small" set of points such that the chord functions at these points distinguish K from all other convex sets.

(iv) Existence. Given a set of points $x_1, \ldots, x_k$ find necessary or sufficient conditions on functions $f_1, \ldots, f_k$ for there to exist a convex set with chord function f_i at x_i $(1 \leq i \leq k)$.

(v) Descriptive. Deduce qualitative properties of a convex set K from properties of its chord functions at a set of points. For instance, if $f_1, \ldots, f_k$ are all r times differentiable, then is the boundary of K an r times differentiable curve?

A good deal of progress has been made on these problems in the last few years, but there are still many open questions. For the point source problem, Falconer has shown, using methods from dynamical systems, that, given points x_1 and x_2 and functions f_1 and f_2, there is at most one convex set with x_1 and x_2 as interior points and chord functions f_1 and f_2 at these points. Moreover, in principle at least, the method is constructive. Similarly, there are at most two such convex sets K with x_1 and x_2 as exterior points and the line through x_1 and x_2 intersecting the interior of K. One possibility has x_1 and x_2 on the same side of K, the other on opposite sides. Surely it should always be possible to eliminate one of these cases, but a complete general argument is not known. In the situations where Falconer's method can be applied, the boundary of K always inherits the order of differentiability of the chord functions.

Necessary and sufficient conditions for a pair of chord functions to correspond to some convex set will be hard to find—a complete characterization would imply a solution to the equichordal problem (see Section A1). However, some (rather weak) necessary conditions are known.

The best result in the general case for the point source problem is due to Volčič. He proves that the chord functions at any four points (no three collinear) serve to distinguish between all convex sets; but the chord functions at any three noncollinear points suffice to distinguish between convex sets not containing the given points. We believe, however, that it ought to be possible to manage with two points in all cases.

The known results and methods in the parallel beam case are completely different. The earliest result is due to Giering who showed that given a convex set K, one can find three directions (depending on K) such that x-ray photographs in these directions are enough to distinguish K from any other convex set. More recently, Gardner & McMullen employed a result of Darboux (see Section B25) on sequences of polygons to show, in an elegant way, that the x-rays in a set of directions distinguish between all convex sets, if and only if the directions are not a subset of the diagonal directions of an affine copy of some regular polygon. Thus, provided these exceptional cases are avoided, four directions suffice. This method is totally nonconstructive—it would be nice to find an algorithm for reconstruction in this situation.

Problems of a different nature arise if the unknown set is known to be a convex n-gon. Edelsbrunner & Skiena show that the (positive) lengths of intersection with $3n + 33$ lines are enough to determine a convex n-gon, given an interior point. (Note that the lines are chosen successively as the measurements are made.) On the other hand, examples show that at least $(3n - 1)/2$ lines are necessary. Exactly how few lines suffice?

One can, of course, generalize these problems to nonconvex sets. In general, a very different approach would be required. Presumably one would need some sort of condition on the sets (e.g., that they are a finite union of convex sets). It does not even seem trivial to exhibit two arbitrary sets that have the same chord functions at two distinct points.

We should mention the idea of x-ray tomography that underlies cancer scanners and which has had a tremendous impact on medical science. For this, we wish to recover a *function* on some (2- or 3-dimensional) domain, given its integrals along certain straight lines. Thus Hammer's x-ray problem is the special case when the function is the characteristic function of a convex set. There is a vast literature on this subject, which belongs to the realms of analysis rather than geometry (e.g., see the AMS Symposium Proceedings edited by Shepp).

G. Bianchi & M. Longinetti, Reconstructing plane sets from projections, *Discrete Comput. Geom.* **5** (1990) 223–242.

H. Edelsbrunner & S. S. Skiena, Probing convex polygons with x-rays, *SIAM J. Comput.* **17** (1988) 870–882; *MR* **89i**:52002.

K. J. Falconer, X-ray problems for point sources. *Proc. London Math. Soc.* (3) **46** (1983) 241–262; *MR* **85g**:52001a.

K. J. Falconer, Hammer's x-ray problems and the stable manifold theorem, *J. London Math. Soc.* (2) **28** (1983) 149–160; *MR* **85g**:52001b.

R. J. Gardner, Symmetrals and x-rays of planar convex bodies, *Arch. Math. (Basel)* **41** (1983) 183–189; *MR* **85c**:52007.

R. J. Gardner, Chord functions of convex bodies, *J. London Math. Soc.* (2) **36** (1987) 314–326; *MR* **88h**:52006.

R. J. Gardner, X-rays of polygons, *Discrete Combin. Geom.*, to appear.

R. J. Gardner & P. McMullen, On Hammer's x-ray problem, *J. London. Math. Soc.* (2) **21** (1980) 171–175; *MR* **81m**:52009.

O. Giering, Bestimmung von Eibereichen und Eikörpen durch Steiner-Symmetrisierungen, *Bayer. Akad. Wiss. Math.-Natur Kl. S.-B.* (1962) 225–253; *MR* **30** #3410.

P. C. Hammer, Problem 2, [Kle], 498–499.

D. Kölzow, A. Kuba & A. Volčič, An algorithm for reconstructing convex bodies from their projections, *Discrete. Comp. Geom.* **4** (1989) 205–237.

M. Longinetti, Some questions of stability in the reconstruction of plane convex sets from projections, *Inverse Problems* **1** (1985) 87–97; *MR* **86f**:52006.

G. Mägerl & A. Volčič, On the well-posedness of the Hammer x-ray problem, *Ann. Mat. Pura Appl.* **144** (1986) 173–182; *MR* **89a**:52006.

L. A. Shepp (ed.), Computed tomography, Proc. Symp. Applied Math. 27, Amer. Math. Soc., Providence, 1983; *MR* **84b**:92012.

A. Volčič, A new proof of the Giering theorem, *Rend. Circ. Math. Palermo* (2) **8** (1985) 281–295; *MR* **88d**:52002.

A. Volčič, A three-point solution to Hammer's x-ray problem, *J. London Math. Soc.* (2) **34** (1986) 349–359; *MR* **87j**:52005.

A. Volčič & T. Zamfirescu, Ghosts are scarce, *J. London Math. Soc.* (2) **40** (1989) 171–178.

A3.   Concurrent normals.

A line is a **normal** of a convex set K in $\mathbb{R}^2$ (resp. $\mathbb{R}^d$) at a boundary point **x** if it is perpendicular to some tangent line [resp. $(d-1)$-dimensional plane] touching K at **x**. An old problem is to study the points of K that lie on several normals. It is conjectured that any convex body in $\mathbb{R}^d$ has an interior point lying on normals through $2d$ distinct boundary points, see Figure A6. Heil has proved this for $d = 2$ and 3, using a combination of Morse theory and approximation. However, in higher dimensions although there is always at least a six normal point, the conjecture remains open and may well require sophisticated techniques from differential topo-

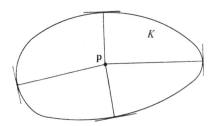

Figure A6.   A convex set K with a "four-normals" point **p**.

logy. If K is centro-symmetric, there are d **double normals** (that is chords that are normal at both ends) through the center, so the conjecture holds in this case (see Lyusternik & Schnirelmann). If K has a twice differentiable boundary, and either the insphere or the circumsphere of K touches at exactly $d + 1$ points, it is obvious that $d + 1$ normals pass through the center of the sphere. However in this situation we do even better, since in fact there must be $2d + 2$ normals through the center.

Zamfirescu has shown, rather surprisingly, that, in the sense of Baire category based on the Hausdorff distance between convex sets, "most" interior points of most convex bodies lie on infinitely many normals. (By most we mean for all but a countable union of nowhere dense sets.) However, we should remember that in the category sense most convex bodies are "exceptional" in that they have highly irregular boundaries that are not even twice differentiable.

For bodies of constant width all normals are double normals. In this case we would expect to find an interior point that lies on normals from $4d - 2$ distinct boundary points; again this has been proved for $d = 2$ and 3.

In the plane, Guggenheimer asks the intriguing question of whether every point in a certain (curvilinear) triangular region must lie on normals from four boundary points.

N. Deo & M. S. Klamkin, Existence of four concurrent normals to a smooth closed curve, *Amer. Math. Monthly* **77** (1970) 1083–1084; *M R* **42** #6718.

H. W. Guggenheimer, Does there exist a "four normals triangle"? *Amer. Math. Monthly* **77** (1970) 177–179.

E. Heil, Concurrent normals and critical points under weak smoothness assumptions, [GLMP] 170–178; *M R* **87d**:52006.

E. Heil, Existenz eines 6-Normalenpunktes in einem konvexen Körper, *Arch. Math. (Basel)* **32** (1979) 412–416, correction **33** (1979) 496; *M R* **80j**:52002.

L. A. Lyusternik & L. G. Schnirelmann, *Méthodes Topologiques dans les Problèmes Variationels*, Hermann, Paris, 1934.

T. Zamfirescu, Intersecting diameters in convex bodies, [RZ], 311–316; *M R* **87a**:52013.

T. Zamfirescu, Points on infinitely many normals to convex surfaces, *J. Reine Angew. Math.* **350** (1984) 183–187; *M R* **85d**:52003.

A4. Billiard ball trajectories in convex regions.

Let K be a plane convex region with boundary curve C. An idealized point "billiard ball" travels across K in a straight line at constant speed and rebounds with equal angles of incidence and reflection on hitting C. Study of billiard ball trajectories involves complex ideas from ergodic theory and dynamical systems; we mention here a few of the more intuitive problems.

It should be noted that some care is required in setting up these problems. Halpern points out that even when C is a three times differentiable curve it is actually possible for the billiard to pass from inside to outside K if the angle of reflexion law is strictly adhered to! This paradox can occur when there are infinitely many bounces in a finite time, but is avoided if the third derivative of C is continuous. The two cases of particular interest are when C is smooth, say infinitely differentiable, and when C is a polygon. In the latter case some convention is required in the exceptional case when a trajectory hits a vertex (e.g., reflection at an equal angle to the angle bisector).

We first look for periodic billiard ball trajectories. Since any smooth plane convex set has at least two distinct **double normals** (i.e., chords cutting C perpendicularly at each end (see Section A3)) there are always two distinct "to and fro" orbits. Steinhaus observed that the inscribed triangle of greatest perimeter in any smooth K gives a periodic triangle, (see Figure A7) and he asked whether there was always a second triangular orbit. This was proved by Croft & Swinnerton-Dyer in an entertaining paper using the "principle of Buridan's ass" (mathematically, that if a dynamical system admits several stable positions of equilibrium, it also admits unstable ones). Similar ideas lead to the conclusion that there are always at least $\varphi(k)$ distinct periodic k-gons, where φ is Euler's totient function (i.e. $\varphi(k)$ is the number of integers no greater than k and coprime to k). Is this true for any number larger than $\varphi(k)$?

The situation is rather different if K is a convex polygon. If K is an acute-angled triangle there is always a periodic triangular orbit (the triangle of least perimeter with one vertex on each side of K). By taking small parallel displacements of this orbit, one obtains an infinite family of periodic hexagons. Further, Mazur has shown that if K is any polygon with vertex angles rational multiples of π then there is a periodic orbit. Other than this, very little is known. Does every convex polygon have a periodic orbit? Or even every obtuse-angled triangle? Are there always orbits of arbitraily large periods?

These problems have obvious analogs for a ball bouncing round a d-dimensional convex body K. Kuiper shows that such K have at least d double normals, giving at least d periodic segments. The work of Croft & Swinnerton–Dyer implies that if K is smooth there are two periodic triangles, but surely there ought to be at least d. Longer orbits seem even harder to analyze.

Steinhaus asked if there are periodic orbits in every polyhedron. Conway has shown that circuits exist in all tetrahedra. He also says that the regular tetrahedron has a continuous family of periodic quadrangles, two periodic hexagons and no other orbits with fewer that eight bounces (see also Gardner). Are there periodic orbits with arbitraily many bounces, and do they all have an even number of bounces?

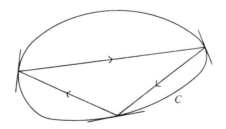

Figure A7.   A period-three billiard trajectory in C.

Another type of billiard problem concerns trajectories that are **dense** in K (that is trajectories that eventually pass arbitrarily close to each point of K). In a rectangle, the trajectory through any given point in any but a countable number of directions is dense—this is a simple consequence of Kronecker's theorem (see, e.g., Hardy & Wright). Zemlyakov & Katok as well as Boldrighini, Keane, & Marchetti show that if K is a convex polygon whose angles are rational multiples of π then the trajectories emanating from all points of K in all but countably many directions are dense in K. For other triangles and polygons it is not even known if there is any dense orbit at all, though it has been shown that most polygons (in the sense of Baire category) have dense orbits.

Many *smooth* convex regions such as the circle or ellipse do not admit any dense orbits. The questions of interest in this situation concern the **caustics** formed by the segments of the orbits. The article by Turner gives a survey of the complex questions that arise.

It is easy to think of poorly posed variants which could be put into a mathematical framework. Several are suggested by W. S. Gilbert in the Mikado's song:

> "On a cloth untrue
> With a twisted cue
> And elliptical billiard balls."

We can pose these problems on *curved* surfaces (for example Robnik & Berry have looked at a problem of this nature involving billiards moving in a magnetic field). What is the effect of an error in the initial direction of travel? What happens if the "ball" is nonspherical? Such questions arise in the study of molecular collisions.

A further type of problem is to characterize convex sets from properties of their billiard trajectories. For example, De Temple & Robertson show that if K is a polygon with a periodic billiard orbit geometrically similar to K then K must be regular, and Sine & Kreĭnović show that a smooth plane convex set K is of constant width if every billiard path has all reflections to the left or all reflections to the right. Similarly, Sine proves that a smooth 3-dimensional convex set is spherical if every path is contained in a plane. Are there further theorems of this nature to do discovered?

Good billard bibliographies are given by Gutkin (for polygons) and Turner (for smooth regions). For related problems in non-convex regions see Section A5.

C. Boldrighini, M. Keane & F. Marchetti, Billiards in polygons, *Ann. Probab.* **6** (1978) 532–540; *MR 58* #31007b.

L. A. Bunimovich, On the ergodic properties of nowhere dispersing billiards, *Commun. Math. Phys.* **65** (1979) 295–312; *MR* **80h**:58037.

H. T. Croft & H. P. F. Swinnerton-Dyer, On the Steinhaus billiard table problem, *Proc. Cambridge Philos. Soc.* **59** (1963) 37–41; *MR* **26** #2925.

D. W. De Temple & J. M. Robertson, A billiard path characterization of regular polygons, *Math. Mag.* **54** (1981) 73–75; *MR* **84g**:52002.

D. W. De Temple & J. M. Robertson, Convex curves with periodic billiard polygons, *Math. Mag.* **58** (1985) 40–42.

M. Gardner, *Martin Gardner's Sixth Book of Mathematical Puzzles and Diversions from the Scientific American*, University of Chicago Press, 1971, Chicago, 29–38.

W. S. Gilbert, The Mikado, Act II, "My object all sublime."

E. Gutkin, Billiards in polygons, *Physica D* **19** (1986) 311–333.

B. Halpern, Strange billiard tables, *Trans. Amer. Math. Soc.* **232** (1977) 297–305; *MR* **56** #9595.

G. H. Hardy & E. M. Wright, *Introduction to the Theory of Numbers*, 4th ed, Oxford University Press, Oxford, 1960, 373.

N. Innami, Convex curves whose points are vertices of billiard triangles, *Kodai Math. J.* **11** (1988) 17–24; *MR* **89d**:52004.

S. Kerckhoff, H. Masur & J. Smillie, A rational billiard flow is uniquely ergodic in almost every direction, *Bull. Amer. Math. Soc.* **13** (1985) 141–142; *MR* **86j**:58080.

N. H. Kuiper, Double normals of convex bodies, *Israel J. Math.* **2** (1964) 71–80; *MR* **30** #4191.

H. Masur, Rational billiards have periodic orbits, to appear.

H. Poritsky, The billiard ball problem on a table with a convex boundary—An illustrative dynamical problem, *Ann. Math.* (2) **51** (1950) 446–470; *MR* **11**, 373.

M. Robnik & M. V. Berry, Classical billiards in magnetic fields, *J. Phys. A* **18** (1985) 1361–1378; *MR* **86j**:58101.

R. Sine, A characterization of the ball in R^3, *Amer. Math. Monthly* **83** (1976) 260–261; *MR* **53** #1404.

R. Sine & V. Kreĭnovič, Remarks on billiards, *Amer. Math. Monthly* **86** (1979) 204–206; *MR* **80k**:28021.

H. Steinhaus, Problems P.175, P.176, P.181, *Colloq. Math.* **4** (1957) 243, 262.

P. H. Turner, Convex caustics for billiards in R^2 and R^3, in [KB], 85–106; *MR* **83i**:52016.

A. N. Zemlyakov & A. B. Katok, Topological transitivity of billiards in polygons, *Mat. Zametki* **18** (1975) 291–300 [translated in *Math. Notes* **18** (1975) 760–764]; *MR* **53** #3267.

A5. Illumination problems. Although this section concerns nonconvex sets, it is so closely related to the previous section that we include it here. We consider point sources of light that emit rays in all directions which are reflected by the mirrored walls of a (2-dimensional) room. If the room is bounded by a re-entrant polygon *C*, can a light source always be positioned so as to illuminate all points in the room after sufficient reflections? More

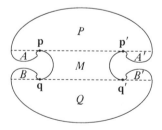

Figure A8.   The upper and lower curves are half-ellipses with foci of **p**, **p′** and **q**, **q′**. Using the fact that an ellipse reflects any ray passing between the foci back between the foci, it is easy to see that rays originating in *P* or *M* can never reach *B* or *B′*.

strongly, can a polygonal room be so illuminated from *any* interior point? Otherwise, is there, for each positive integer k, a polygonal room not illuminable by any k lights?

For *non*-polygonal regions, these questions have negative answers. Various people developed an idea of Penrose & Penrose to get a smooth region not illuminable from any of its points: the upper and lower portions of the boundary are ellipses with foci at the dark points. The exact shape of the central portion of the boundary is immaterial. Rauch gives a number of examples that throw further light on the subject. He constructs a region smooth (i.e., continuously differentiable) at all but one of its boundary points that requires infinitely many sources for its illumination. He points out that if C is *everywhere* smooth then one can always manage with finitely many sources, but gives examples requiring at least k sources for every k.

Pach poses the following very elegant question: if you light a match in a forest of reflecting trees, can the light be seen outside? Formally, if $K_1, \ldots, K_n$ are disjoint closed disks whose boundaries reflect light, is there a light path from every point of $\mathbb{R}^2 \setminus \bigcup_{i=1}^{n} K_i$ out to infinity? What if the K_i are arbitrary smooth convex sets? A solution to these problems might have applications in acoustics.

Another question on reflections is asked by Wunderlich. Does there exist a closed analytic or even algebraic curve such that any light ray from one "focus" **x** always passes through another focus **y** after exactly two reflections. (If **x** = **y**, the answer is yes, and of course an ellipse with **x** and **y** as foci gives the affirmative answer if we demand one reflection rather than two.)

R. K. Guy & V. Klee, Monthly research problems, *Amer. Math. Monthly* **78** (1971) 1114.

V. Klee, Some unsolved problems in plane geometry, *Math. Mag.* **52** (1979) 131–145; *MR* **80m**:52006.

V. Klee, Is every polygonal region illuminable from some point? *Amer. Math. Monthly* **76** (1969) 180.

L. Penrose & R. Penrose, Puzzles for Christmas, *New Scientist*, 25 Dec. 1958, 1580–1581.

J. Rauch, Illuminations of bounded domains, *Amer. Math. Monthly* **85** (1978) 359–361.

W. Wunderlich, Ein Spiegelproblem, *Monatsch. Math.* **53** (1949) 63–72.

W. Wunderlich, Ungelöste Probleme 35, *Elem. Math.* **15** (1960) 37–39.

A6. The floating body problem. Ulam proposed the following intriguing problem. If a convex body K made of a material of uniform density $\rho < 1$ floats in equilibrium in any orientation (in water, of density 1), must K be spherical?

This is true if $\rho = \frac{1}{2}$ and K is centro-symmetric (see Schneider or Falconer). It is also true in the limiting case where $\rho = 0$ (that is for a body that can rest on a plane table in any orientation). As Montejano points out, any section through the center of gravity **g** has $dr/d\theta = 0$ with respect to **g** as origin for polar coordinates, so each such section is circular and K must be spherical. If a body K floats in any orientation in liquids of all sufficiently large densities

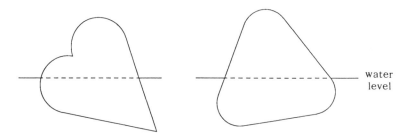

Figure A9.   Sections of "logs" of density 1/2 that will float in any orientation.

then K is spherical; this follows by letting $\rho \to 0$ and using the above. Are two distinct densities, or even a finite number, enough to ensure sphericity?

The "2-dimensional" version of the problem, for a floating log of uniform cross-section, is also intriguing. In this case, of course, we seek a log that will float in every orientation with its axis horizontal. Auerbach has exhibited such logs with density $\rho = \frac{1}{2}$ both with nonconvex and convex but noncircular sections (see Figure A9). However, if $\rho \neq \frac{1}{2}$, no progress has been made.

One can concoct a range of 3-dimensional problems as follows: Denote plane sections of a convex body K by letters:

V: those that cut off a given constant volume;
A: those with the section having a certain constant area;
S: those that cut off a given surface area;
B: those with boundary curve of given perimeter; and
I: those with plane section having equal principal moments of inertia.

We may then take ordered pairs of letters to denote problems in the following way: (V, S): if all sections of type V are also type S, then is K a sphere? (In some special instances, such as when the constant volume and surface area are half those of K, the pertinent question is whether K is centro-symmetric.) It seems that the floating body problem is just problem (V, I).

H. Auerbach, Sur un problème de M. Ulam concernant l'équilibre des corps flottants, *Studia Math.* **7** (1938) 121–142.

K. J. Falconer, Applications of a result on spherical integration to the theory of convex sets, *Amer. Math. Monthly* **90** (1983) 690–693; *MR* **85f**:52012.

R. D. Mauldin, [Mau], Problem 19.

L. Montejano, On a problem of Ulam concerning a characterization of the sphere, *Studies Appl. Math.* **53** (1974) 243–248; *MR* **50** #8296.

R. Schneider, Functional equations connected with rotations and their geometric applications, *L'Enseign. Math.* **16** (1970) 297–305; *MR* **44** #4642.

S. M. Ulam, *A Collection of Mathematical Problems*, Interscience, New York, 1960, p. 38; *MR* **22** #10884.

A7.   Division of convex bodies by lines or planes through a point. Suppose that K is a convex set in $\mathbb{R}^3$ and $\mathbf{x}$ is an interior point of K such that any plane

through **x** divides the surface of K into two parts of equal surface area. Must K be centro-symmetric about **x**?

It is not hard to prove the two-dimensional analog of this, that if every chord through **x** of a plane convex set K divides the perimeter into parts of equal length then K has central symmetry about **x**. The interesting problem in the plane case is whether every convex K has an interior point **x** such that every chord through **x** divides the perimeter into two parts, the ratio of whose lengths is always less than the golden ratio $\frac{1}{2}(\sqrt{5} + 1)$; this was conjectured by Neumann. Consideration of very thin isosceles triangles shows that the figure $\frac{1}{2}(\sqrt{5} + 1)$ cannot be replaced by a smaller ratio.

The analogs of these questions, but considering the area of K on either side of a chord and, in $\mathbb{R}^3$, the volume of K on either side of a plane, have been completely settled, and are chronicled in Grünbaum's article in connexion with Winternitz's measure of symmetry.

B. Grünbaum, Measures of symmetry for convex sets, in [Kle], 233–270; *MR* **27** #6187.

B. H. Neumann, On an invariant of plane regions and mass distributions, *J. London Math. Soc.* **20** (1945) 226–237; *MR* **8**, 170.

A8. Sections through the centroid of a convex body. Let K be a 3-dimensional convex body with centroid (i.e., center of gravity) **g**. Is **g** necessarily the centroid of at least four plane sections of K through **g**? Is it even the centroid of seven such sections, as is the case if K is a tetrahedron? More generally, if K is a d-dimensional convex body, is the centroid of K the centroid of $d + 1$ or even of $2^d - 1$ of the $(d - 1)$-dimensional sections through **g**? When $d = 2$ this is easily seen to be so—in this case **g** bisects three chords of K. This question is due to Grünbaum and Loewner, see also the earlier paper by Steinhaus.

A consequence of Helly's theorem (see Section E1) is that *some* point of K is the centroid of at least $d + 1$ sections by hyperplanes. What can be said about the set of points of K enjoying this property? In the plane case Ceder showed that this set is connected, but not necessarily convex. Chakerian & Stein discuss other aspects of this problem.

J. G. Ceder, On a problem of Grünbaum, *Proc. Amer. Math. Soc.* **16** (1965) 188–189; *MR* **30** #3413.

G. D. Chakerian & S. K. Stein, Bisected chords of a convex body, *Arch. Math. (Basel)* **17** (1966) 561–565; *MR* **34** #6635.

B. Grünbaum, On some properties of convex sets, *Colloq. Math.* **8** (1961) 39–42; *MR* **23** #A3508.

B. Grünbaum, Measures of symmetry for convex sets, in [Kle], 233–270; *MR* **27** #6187.

C. Loewner, Problem 28, in [Fen], 323.

H. Steinhaus, Quelques applications des principes topologiques à la géométrie des corps convexes, *Fund. Math.* **41** (1955) 284–290; *MR* **16**, 849.

A9. **Sections of centro-symmetric convex bodies.** Let K and J be convex bodies in $\mathbb{R}^3$, both centro-symmetric about the origin **o**. Suppose that $A(K \cap P) \le A(J \cap P)$ for every plane P through **o**. Does this imply the volume inequality $V(K) \le V(J)$? This deceptively simple-sounding problem is due to Busemann & Petty. Busemann shows that the result is certainly false if the centro-symmetric stipulation is dropped or if "convex" is replaced by "star-shaped about **o**" (i.e., with each point of K joined to **o** by a line segment in K). Giertz shows that it is true under additional symmetry considerations.

The first significant progress was made by Larman & Rogers who used a remarkable probabilistic construction to show that the d-dimensional analog of the conjecture is false if $d \ge 12$. They exhibit a centro-symmetric K in $\mathbb{R}^d$ ($d \ge 12$) with $V_{d-1}(K \cap P) \le V_{d-1}(J \cap P)$ for every hyperplane P, but with $V_d(K) > V_d(J)$, where J is the unit ball. (V_d denotes d-dimensional volume). However, the detailed calculations fail for smaller d. More recently, Ball showed that the largest $(d-1)$-dimensional section of the unit cube in $\mathbb{R}^d$ has $(d-1)$-dimensional volume not exceeding $\sqrt{2}$—a nontrival calculation. However, if B is the ball of unit d-dimensional volume in $\mathbb{R}^d$, then $V_{d-1}(B \cap P) > \sqrt{2}$ for any plane P through the center of B, provided that $d \ge 10$. Thus by taking K as the unit cube and J as a slightly reduced copy of B, the d-dimensional analog of the conjecture fails if $d \ge 10$. Very recently, Giannopoulos and Bourgain have shown that it fails if $d \ge 7$. Are there counter-examples in fewer dimensions?

The same question, but with sections replaced by projections was answered in the negative by Schneider. Here K and J are centro-symmetric with $V_{d-1}(\text{proj}_P K) \le V_{d-1}(\text{proj}_P J)$ for all hyperplanes P, where proj_P denotes orthogonal projection onto P. It is immediate from Cauchy's surface area formula that the $(d-1)$-dimensional surface area of K is no more than that of J, but it is not in general true that $V_d(K) \le V_d(J)$, even for $d = 3$.

If K and J are centro-symmetric and every affine image of K has $(d-1)$-dimensional surface area at most equal to that of the corresponding affine image of J, does it follow that $V_d(K) \le V_d(J)$? Fáry & Makai point out that it does if $d = 2$.

K. Ball, Cube slicing in $\mathbb{R}^n$, *Proc. Amer. Math. Soc.* **97** (1986) 465–473; *MR* **87g**:60018.

K. Ball, Some remarks on the geometry of convex sets, in *Geometric Aspects of Functional Analysis*, Springer Lecture Notes in Math, 1317, Springer, Berlin, 1988, 224–231; *MR* **89h**:52009.

J. Bourgain, On the Busemann–Petty problem for perturbations of the ball, *Geom. Func. Analysis* **1** (1991) 1–13.

H. Busemann, Volumes and areas of cross sections, *Amer. Math. Monthly* **67** (1960) 248–250, correction **67** (1960) 671; *MR* **22** #11313, **23** #A3826.

H. Busemann & C. M. Petty, Problem 1, Problems on convex bodies, *Math. Scand.* **4** (1956) 88–94; *MR* **18**, 922.

I. Fáry & E. Makai, Research problems, *Period. Math. Hungar.* **14** (1983) 111–114.

I. Fáry & E. Makai, Problem, in [BF], 694–695.

A. A. Giannopoulos, A note on a problem of H. Busemann and C. M. Petty concerning sections of convex bodies, *Mathematika* **37** (1990) 239–244.

M. Gietz, A note on a problem of Busemann, *Math. Scand.* **25** (1969) 145–148; *MR* **41** #7534.

E. L. Grinberg & I. Riven, Infinitesimal aspects of the Busemann–Petty problem, *J. London Math. Soc.*, to appear.

D. G. Larman & C. A. Rogers, The existence of a centrally symmetric convex body with central sections that are unexpectedly small, *Mathematika* **22** (1975) 164–175; *MR* **52** #11737.

R. Schneider, Zu einem Problem von Shephard über die Projektionen konvexer Körper, *Math. Z.* **101** (1967) 71–82; *MR* **36** #2059.

A10. **What can you tell about a convex body from its shadows?** Let K be a convex body in 3-space, and let K_P be the orthogonal projection or "shadow" of K on the plane P. What can be deduced about K from some knowledge of the K_P? The area $A(K_P)$ is sometimes known as the **outer quermass** or **brightness** of K relative to P [see Figure A10(a)]. The surface area of K may be found from the brightnesses using Cauchy's surface area formula, $S = \frac{1}{\pi} \int A(K_P)\, dP$, where the integral is over all planes through the origin and with respect to solid angle. However, K is not even determined up to translation and reflection in a point by a knowledge of $A(K_P)$ for all P. Many years ago Blaschke and Bonnesen constructed nonspherical bodies of constant brightness [i.e., with $A(K_P)$ constant for all P], and higher dimensional analogs have been provided by Firey. The outstanding unsolved problem is whether there exists a nonspherical body in $\mathbb{R}^3$ of both constant brightness *and* constant width (i.e., $w(\theta)$ constant for all θ). Such a body cannot have a smooth (twice differentiable) surface—see Bonnesen & Fenchel for Matsumura's proof of this.

The **isoperimetric quotient** $4\pi A(K_P)/L(K_P)^2$ is a measure of the "circularity" of the projection K_P. Chakerian asks whether K must be spherical if $A(K_P)/L(K_P)^2$ is independent of P— this would imply that bodies of both constant brightness and constant width are indeed spherical.

Santaló asked us whether *some* shadow always has a relatively short perimeter, that is whether

$$\min_P L(K_P)^2 \le \left(2 + \frac{4}{\sqrt{3}}\right) S,$$

with equality just for the regular tetrahedron, where S is the surface area of K.

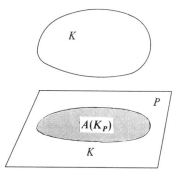

Figure A10(a).   The area of projection of K on plane P is the outer quermass $A(K_P)$ of K.

W. Blaschke, [Bla], 140.

T. Bonnesen, Om Minkowski's uligheder fur konvexer legemer, *Mat. Tidsskr.* **B** (1926) 80.

T. Bonnesen & W. Fenchel, [Bon Fe], 140.

G. D. Chakerian, Is a body spherical if all its projections have the same I.Q.? *Amer. Math. Monthly* **77** (1970) 989–992.

W. J. Firey, Blaschke sum of convex bodies and mixed bodies, in [Fen], 94–101; *MR* **36** #3231.

A11. **What can you tell about a convex body from its sections?** Some of these problems are, in a sense, dual to those of the previous section.

Let E be a 3-dimensional object. For each plane P write $Q_P(E)$ for the largest area of the intersection of E with a plane that is parallel to P. Then $Q_P(E)$ is called the **inner quermass** or the **HA measurement** of E relative to P [see Figure A10(b)]. Under what circumstances can E be determined if $Q_P(E)$ is given for all P? This problem probably originates with Bonnesen, but more recently it has found important physical applications in the study of the Fermi surfaces of metals. One wishes to determine the surface from the HA-measurements, which can be found by means of the de Haas–van Alphen effect. For more detailed accounts of the physics involved see Mackintosh or Shoenberg.

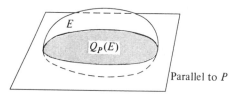

Figure A10(b). The maximum area of intersection of E with a plane parallel to P, gives the inner quermass $Q_P(E)$ of E.

The nature of these problems depends to a large extent on the class of E considered. Zaks described a large family of "hollow" sets for which $Q_P(E)$ is constant for all planes P. His examples are based on the well-known fact that if B_a and B_b are concentric balls of radii a and b $(a < b)$ then any plane cutting B_a intersects the region $B_b \backslash B_a$ in area $\pi(b^2 - a^2)$. If not too much of $B_b \backslash B_a$ is removed, one still has an object with $Q_P = \pi(b^2 - a^2)$ for all P.

Thus the problems of interest concern convex objects. Is a convex body K uniquely determined if $Q_P(K)$ is given for all P, and in particular is K spherical if $Q_P(K)$ is constant? If K is centro-symmetric, then the answers to these questions are affirmative, see Funk or Lifschitz & Pogorelov or Falconer. Bonnesen's original question was whether K is determined if both the inner *and* outer quermasses $Q_P(K)$ and $A(K_P)$ are specified (see Section A10). Even this remains unsolved, though Zaks's examples show that the convexity condition cannot be dispensed with.

T. Bonnesen, Om Minkowski's uligheder for konvekse legemer, *Mat. Tidsskr* **B** (1926) 80.

T. Bonnesen & W. Fenchel, [Bon Fe], 140.

K. Falconer, Applications of a result on spherical integration to the theory of convex sets, *Amer. Math. Monthly* **90** (1983) 690–693; *M R* **85f**:52012.

W. Fenchel, To problem 22, [Fen], 322.

P. Funk, Über Flächen mit lauter geschlossenen geodätischer Linien, *Math. Ann.* **74** (1913) 278–300.

V. Klee, Is a body spherical if all its HA measurements are constant? *Amer. Math. Monthly* **76** (1969) 539–542.

V. Klee, Problem 22, [Fen], 321.

I. M. Lifschitz & A. V. Pogorelov, On the determination of Fermi surfaces and electron velocities in metals by the oscillation of magnetic susceptibility, *Dokl. Akad. Nauk SSSR* **96** (1954) 1143–1145.

A. R. Mackintosh, The Fermi surface of metals, *Scientific American*, **209**, no. 1. (July 1963), 110–120.

D. Shoenberg, The de Haas–von Alphen effect, in *The Fermi Surface*, W. A. Harrison and M. B. Webb (eds.), Wiley, New York, 1960, 74–83.

J. Zaks, Nonspherical bodies with constant HA measurements exist, *Amer. Math. Monthly* **78** (1971) 513–516.

A12.   Overlapping convex bodies. Fickett proposed the following neat problem: let C and C' be the perimeters of overlapping congruent rectangles K and K'. Is it true that for any relative positions

$$\tfrac{1}{3} \leq \text{length } (C \cap K')/\text{length } (C' \cap K) \leq 3?$$

(See Figure A11.) This ratio certainly lies between $\tfrac{1}{4}$ and 4. More generally, what are the bounds of this ratio for other congruent convex sets K and K' with bounding curves C and C'? In the case of congruent disks it is identically one. If C and C' are congruent triangles, is the maximum cosec $\tfrac{1}{2}\theta$ where θ is the least angle?

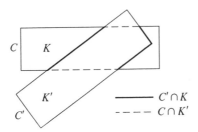

Figure A11.   How small or large can the ratio of the length of $C' \cap K$ to that of $C \cap K'$ be?

If C and C' are d-dimensional rectangular parallelepipeds, is the analogous upper bound, with length replaced by "$(d - 1)$-dimensional surface area", equal to $2d - 1$?

J. W. Fickett, Overlapping congruent convex bodies, *Amer. Math. Monthly* **87** (1980) 814–815.

A13. **Intersections of congruent surfaces.** Must the surfaces of three congruent centro-symmetric convex bodies in $\mathbb{R}^3$ with common center have a point in common? This is true for bodies that are axially-symmetric as well as centro-symmetric (see Hadwiger).

Let C be the surface of a 3-dimensional convex body. Suppose that $C \cap C'$ is either empty or a connected set for every congruent copy C' of C. Schäffer asks whether C is necessarily a "sausage" [i.e., a circular cylinder with two hemispherical caps (with the sphere as a special case)].

H. Hadwiger, Ungeloste Probleme 50, *Elem. Math.* **23** (1968) 90.
J. J. Schäffer, Problem 11, Durham Symposium on convexity, *Bull. London Math. Soc.* **8** (1976) 28–33.

A14. **Rotating polyhedra.** What closed convex 3-dimensional surfaces enjoy the property that a regular tetrahedron can rotate to any orientation within them, keeping all four vertices always on the surface?

Within a symmetrical lens-shaped surface, formed by two spherical caps each of semiangle $\pi/6$, a regular tetrahedron can move around fairly freely to many orientations, but *not* to all. It may be easier to prove that the surface must be a sphere if we add some smoothness conditions, differentiability everywhere on the surface being the obvious one. In two dimensions, curves inside which an equilateral triangle can slide round can be obtained by slightly deforming a circle near three equidistant boundary points. There is also an analog of the lens.

We can ask the same question for other polyhedra. Non-regular tetrahedra, and the Platonic solids are of particular interest.

These problems seem intrinsically harder than the dual problems of finding rotors in polyhedra. A convex body is a **rotor** of a polyhedron if it can be rotated inside the polyhedron to every orientation with every face of the polyhedron touching the body. For example, any body of constant width is a rotor of the cube. However, there are also polyhedra that have rotors that are not of constant width. For general discussion of rotors see Goldberg or Yaglom & Boltyanskii or Chakerian & Groemer. Meisener showed that the cube, tetrahedron and octahedron are the only bodies in $\mathbb{R}^3$ that can have nonspherical rotors, and Schneider obtained a characterization in $\mathbb{R}^d$.

G. D. Chakerian & H. Groemer, Convex bodies of constant width, in [GW], 49–96; MR **85f**:52001.
H. T. Croft, Some geometrical thoughts, *Math. Gaz.* **49** (1965) 45–49.
H. T. Croft, Convex curves in which a triangle can rotate, *Math. Proc. Cambridge Philos. Soc.* **88** (1980) 385–393; *MR* **82c**:52003.
M. Goldberg, Rotors in polygons and polyhedra, *Math. Comput.* **14** (1960) 229–239; *MR* **22** #5934.
E. Meissner, Uber die durch reguläre Polyeder nicht stützbaren Körper, *Vier. Natur Ges. Zürich* **63** (1918) 544–551.
R. Schneider, Gleitkörper in konvexen Polytopen, *J. Reine Angew. Math.* **248** (1971) 193–220; *MR* **43** #5411.
I. M. Yaglom & V. G. Boltyanskii, *Convex Figures*, Holt, Rinehart, and Winston, New York, 1961; *MR* **23A** #1283.

A15. **Inscribed and circumscribed centro-symmetric bodies.** Let K be a convex body in $\mathbb{R}^d$, let K_I be the largest centro-symmetric convex body inside K and let K_C be the smallest centro-symmetric body containing K (in the sense of area if $d = 2$, volume if $d = 3$, etc.).

It is well-known that for a plane set K:

$$\tfrac{2}{3} \le A(K_I)/A(K) \le 1,$$

with equality on the left if K is a triangle and on the right if K is centro-symmetric. The 3-dimensional analogs are surprisingly awkward: does a convex body contain a symmetric one of half its volume? i.e., is

$$\tfrac{1}{2} \le V(K_I)/V(K) \le 1$$

with equality for the tetrahedron? The best known constant on the left is $\tfrac{2}{9}$, due to Bielecki & Radziszewski. Perhaps this is even true for a K_I with center coinciding with the center of gravity of K. The 2-dimensional result was originally proved by Besicovitch and the centre of gravity analog by Ehrhart, Stewart and Kozinec.

Similarly, it is known that, for a plane set K the following is true:

$$\tfrac{1}{2} \le A(K)/A(K_C) \le 1,$$

again with equality on the left for K a triangle and on the right for K centro-symmetric (see Estermann, Levi, and Fáry). In $\mathbb{R}^3$, is the minimum value of $V(K)/V(K_C)$ equal to $\tfrac{1}{3}$ with the tetrahedron the extremal body? The minimum is known to be at least $\tfrac{1}{6}$.

If K is restricted to plane sets of constant width, then both $A(K_I)/A(K)$ and $A(K)/A(K_C)$ are least when K is a Reuleaux triangle and greatest when K is circular (see Besicovitch, Eggleston, and Chakerian).

The analogs of all these problems in three or more dimensions are wide open, though Chakerian and Chakerian & Stein give some rough lower bounds.

We can ask similar questions, replacing areas or volumes by perimeter lengths or surface areas. Practically nothing seems to be known, even in the plane case (see Grünbaum).

The quantities $A(K_I)/A(K)$ and $A(K)/A(K_C)$, etc., are examples of what Grünbaum calls **measures of symmetry**—they take a maximum value of 1, if and only if K is centro-symmetric, and indicate, in some sense, how nearly symmetrical K is. In particular, $A(K_I)/A(K)$ is called the Kövner–Besicovitch measure of symmetry and $A(K)/A(K_C)$ is Estermann's measure of symmetry. Grünbaum's article contains much more on these and other symmetry measures, and has a large bibliography.

A. S. Besicovitch, Asymmetry of a plane convex set with respect to its centroid, *Pacific J. Math.* **8** (1958) 335–337.

A. S. Besicovitch, Measures of asymmetry of convex curves, II: Curves of constant width, *J. London Math. Soc.* **26** (1951) 81–93; *MR* **12**, 850.

A. Bielecki & K. Radziszewski, Sur les parallélépipèdes inscrits dans les corps convexes, *Ann. Univ. Marie Curie-Sklodowska Sect. A* **8** (1954) 97–100; *MR* **18**, 412.

G. D. Chakerian, Sets of constant width, *Pacific J. Math.* **19** (1966) 13–21; *MR* **34** #4986.

G. D. Chakerian & S. K. Stein, On measures of symmetry of convex bodies, *Canad. J. Math.* **17** (1965) 497–504; *MR* **31** #1612.

G. D. Chakerian & S. K. Stein, On the symmetry of convex bodies, *Bull. Amer. Math. Soc.* **70** (1964) 594–595; *MR* **29** #517.

H. G. Eggleston, Measures of asymmetry of convex curves of constant width and restricted radii of curvature, *Quart. J. Math. Oxford* (2) **3** (1952) 63–72; *MR* **13**, 768.

E. Ehrhart, Propriétés arithmo-géométriques des ovales, *C. R. Acad. Sci. Paris* **241** (1955) 274–276; *MR* **17**, 350.

T. Estermann, Ueber den Vektorenbereich eines konvexen Körpers, *Math. Z.* **28** (1928) 471–475.

I. Fáry, Sur la densité des réseaux de domaines convexes, *Bull. Soc. Math. France* **78** (1950) 152–161; *MR* **12**, 526.

B. Grünbaum, Measures of symmetry for convex sets, in [Kle], 233–270; *MR* **27** #6187.

B. N. Kozinec, On the area of the kernel of an oval, *Leningrad. Gos. Univ. Ui. Zap. Ser. Math. Nauk.* 1958, 83–89.

F. W. Levi, Über zwei Sätze von Herrn Besicovitch, *Arch. Math.* **3** (1952) 125–129; *MR* **14**, 309.

B. M. Stewart, Asymmetry of a plane convex set with respect to its centroid, *Pacific J. Math.* **8** (1958) 335–337; *MR* **20** #4238.

A16.   Inscribed affine copies of convex bodies.

Let J and K be plane convex sets, and let J_0 be the affine copy of J of largest area contained in K. Given K, what is the minimum of $A(J_0)/A(K)$ over all convex J, and given J, what is the minimum of $A(J_0)/A(K)$ over all convex K? Apslund considers this problem when one of the sets is a parallelogram or ellipse, and the other is centro-symmetric. He conjectures that if K and J are both centro-symmetric, there is always some affine copy J_0 of J which, when expanded by a linear factor $\frac{3}{2}$, can cover K.

The most interesting problem is to find the minimum of $A(J_0)/A(K)$ as J and K both range through all plane convex sets. Thomas shows that for any J and K we have $A(J_0)/A(K) \geq \frac{1}{6}$, with $V(J_0)/V(K) \geq 2^d(2!)^2/d^d(2d)!$ for the analogous d-dimensional problem. He conjectures that in the plane case "$\frac{1}{6}$" can be replaced by "$\frac{1}{4}$" and, in general, $V(J_0)/V(K) \geq d^{-d}$, perhaps even with some translate of $-\frac{1}{d}J$ always lying in K. (Here $-\frac{1}{d}J$ is the reflection of J in the origin, scaled by a factor $\frac{1}{d}$.)

E. Asplund, Comparison between plane symmetric convex bodies and parallelograms, *Math. Scand.* **8** (1960) 171–180; *MR* **23** #A2796.

F. Behrend, Über einige Affininvorienten, *Math. Annalen* **113** (1937) 713–747.

F. John, Extremum problems with inequalities as subsidiary conditions, Studies and Essays presented to R. Courant, New York, 1948, 187–204.

R. H. K. Thomas, Pairs of convex bodies, *J. London Math. Soc.* (2) **19** (1979) 144–146; *MR* **80m**:52002.

A17.   Isoperimetric inequalities and extremal problems.

The next few sections are concerned with inequalities relating the various measures associated with convex sets, such as area and perimeter length. The well-known fact that among all plane regions of a given area the circle has the smallest perimeter length, may be expressed as the basic **isoperimetric inequality**

$$L^2 \geq 4\pi A,$$

where L is the length of the boundary and A the area of a convex domain K. Equality holds just for the circle. Similarly, for 3-dimensional convex sets

$$S^3 \geq 36\pi V^2,$$

where S is surface area and V is volume, with equality only for the sphere. These are the prototypes for a vast class of inequalities between the basic measures for convex sets—A, L, (or V, S, M) w, D, r, R, (see the introduction to this chapter). We can also consider physical constants, such as moment of inertia about the center of gravity I, electrostatic capacity C, and the principal frequency f of a vibrating membrane stretched across the region. The aim is to find inequalities, and ultimately a set of best possible inequalities, between the various parameters, and to identify the **extremal sets** for which equality occurs. Such inequalities have important applications in a wide variety of areas of mathematics and science, such as potential theory, differential equations, and optimization theory. Also of interest are inequalities for certain restricted classes of sets, such as polygons or sets of constant width, and, where appropriate, for non-convex sets. A further major area of study concerns isoperimetric inequalities on curved surfaces.

   We can only include a small sample of the problems on this subject. Listed below are some principal references which contain many known and conjectured inequalities of isoperimetric type (see in particular page 249 of the book by Pólya & Szegö for "seventy different conjectures.") Most of the standard references for convexity also treat this subject.

C. Bandle, *Isoperimetric Inequalities and Applications*, Pitman, Boston and London, 1980; *MR* **81e**:35095.

C. Bandle, Isoperimetric inequalities, in [GW], 30–48; *MR* **85f**:52029.

R. Osserman, The isoperimetric inequality, *Bull. Amer. Math. Soc.* **84** (1978) 1182–1238; *MR* **58** #18161.

R. Osserman, Bonnesen-style isoperimetric inequalities, *Amer. Math. Monthly* **86** (1979) 1–29; *MR* **80h**:52013.

L. E. Payne, Isoperimetric inequalities and their applications, *SIAM Review* **9** (1967) 453–488; *MR* **36** #2058.

C. M. Petty, Isoperimetric problems, in [Kay], 26–41; *MR* **50** #14499.

G. Pólya & G. Szegö, *Isoperimetric Inequalities in Mathematical Physics*, Princeton University Press, Princeton, 1951; *MR* **13**, 270.

A18. **Volume against width.** Many years ago, Pál showed that the plane set of width 1 with smallest area is an equilateral triangle. What is the 3-dimensional body of width 1 that has least volume? Equivalently, what is the body of minimal volume that contains a unit segment in each direction? For many years, it was thought that the regular tetrahedron of unit height, with volume 0.333..., was minimal. However, Heil has recently introduced the following set K: Take the regular tetrahedron of edge $\sqrt{2}$ and replace each edge by a circular arc of radius 1 centered in the middle of the opposite edge. Take the four points at distance 1 from the faces of the tetrahedron which lie between a vertex and the center of the opposite face. Let K be the convex hull of these four points and six circular arcs. Heil suggests that K has width 1 and volume 0.298.... Is this the body of width 1 of least volume?

Firey showed that any convex set in $\mathbb{R}^d$ of width 1 has d-dimensional volume at least $2/d! \sqrt{3}$; can better estimates be obtained?

Similarly, what convex body in $\mathbb{R}^3$ of width 1 has the minimum surface area? Is it also the set K described above, with surface area 2.931...?

W. J. Firey, Lower bounds for volumes of convex bodies, *Arch. Math. (Basel)* **16** (1965) 69–74; *MR* **31** #5152.
E. Heil, Comment on Problem 26, [TW], 261.
J. Pál, Ein minimumproblem für Ovale, *Math. Ann.* **83** (1921) 311–319.

A19. Extremal problems for elongated sets.

Bieri poses the following problem for sets constrained to be "long and thin": Given the diameter and width of a plane convex set, what is the minimal possible perimeter length? It is conjectured that the extremal sets are polygons inscribed in a Reuleaux triangle. The other extremal problems of this type (maximum perimeter, minimum and maximum area) all seem to be susceptible to routine methods.

Bieri further asks for a description of the plane convex sets of given diameter, width and perimeter that have maximal, and also that have minimal, area.

Weiss discusses the convex figures of given area and perimeter that have maximum and minimum diameter. He identifies the maximal sets, but leaves open the minimal problem.

H. Bieri, Problems 5 and 6 in [TW], 256.
M. L. Weiss, Convex figures with given area and circumference, and maximal or minimal diameter, *Normat* **32** (1984) 128–134; *MR* **86b**:52006.

A20. Dido's Problem.

> Devenere locos ubi nunc ingentia cernis
> Moenia surgentemque norae Carthaginis arcem,
> Mercatique solum, facti de nomine Bursam,
> Taurino quantum possent circumdare tergo.
>
> Virgilius, Aeneid I, 365–368.

The story is told of how the Queen Dido of Carthage, a refugee from Tyria, asked an African chieftain for as much land as she could enclose with the hide of a cow. She cut the hide into small strips which she laid out in a semicircle, with the North African coast as the other boundary, in order to enclose as great an area as possible.

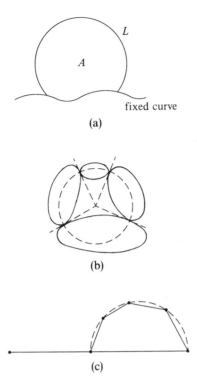

Figure A12. Variants on Dido's problem. (a) The maximum area enclosed by a fixed curve and a curve of given length is achieved by a circular arc. (b) Likely arrangement of smooth convex sets that enclose the greatest area. (c) The maximum area enclosed by "hinged" line segments is attained by a semicircular polygon if one segment is longer than the sum of the lengths of the rest.

This suggests a number of isoperimetric-type problems. What is the greatest area that can be enclosed by a curve of length L using a fixed curve as part of the boundary? The extremal is always given by some circular arc [Figure A12(a)]—this is an easy consequence of the basic isoperimetric property.

Suppose that we are given a collection of convex regions and wish to arrange them to enclose maximum area. Fejes Tóth conjectured that in any maximal arrangement no region comes into contact with more than two others. Assuming this he proves that if the regions are smooth, the points of contact must all lie on a circle, and also that any half-line emanating from the center of this circle can pass through the interior of at most one of the sets [see Figure A12(b)]. However, it seems awkward in general to exclude the possibility of some of the regions crossing.

Fejes Tóth & Heppes have studied the case when the bounding regions are all *translates* of a given convex domain. They conjecture that $n > 2$ translates of a convex domain of area 1 can enclose an area of at most $\lfloor n(n-3)/6 \rfloor + 1$, with equality for suitable translates of a triangle.

How should one arrange n straight-line segments to enclose the maximum area? Presumably the extremal is given by a cyclic polygon. However, it seems hard to show that the segments do not cross. If one of the segments is longer that the sum of the rest, Straus showed that the extremal is given by a "semicircular polygon" [see Figure A12(c)].

The 3-dimensional analog of this presents a good challenge. Given a number of 2-dimensional convex regions, arrange them in space (intersections allowed) to enclose the largest possible volume. Alternatively cut a polygonal "face" out of each region and arrange these to form a polyhedron of maximum volume. What can be said about the extremal configurations? In particular, what happens if we are given n equal spheres?

A. M. Amilibia & S. S. Gomis, A Dido problem for domains in R^2 with a given inradius, *Geom. Dedicata* **34** (1990) 113–124.

A. Bezdek & K. Bezdek, On a discrete Dido-type question, *Elem. Math.* **44** (1989) 92–100; *M R* **90j**:52013.

K. Bezdek, On a Dido-type question, *Ann. Univ. Sci. Budapest. Eötvös Sect. Math.* **29** (1986) 241–244; *M R* **88h**:52017.

L. Fejes Tóth, Über das Didosche Problem, *Elem. Math.* **23** (1968) 97–101; *M R* **38** #5114.

L. Fejes Tóth, On the Dido problem of three discs, *Mat. Lapok* **19** (1968) 9–12; *M R* **41** #924.

L. Fejes Tóth & A. Heppes, Regions enclosed by convex domains, *Studia Sci. Math. Hungar.* **1** (1966) 413–417; *M R* **34** #3432.

A21. Blaschke's Problem.

Find the best possible set of inequalities relating the volume V, surface area S, and integral mean curvature M of a convex body K in space. In other words, exactly what combinations of V, S, and M are simultaneously realizable by a convex body? By virtue of the scaling properties of these quantities under similarity transformations, it is enough to write $x = 4\pi S/M^2$, $y = 48\pi^2 V/M^3$ and to ask which pairs (x, y) can occur (see Figure A13). Estimates using the Brunn–Minkowski inequality easily give that $0 \le x \le 1$ and $0 \le y \le 1$ and that $y \le x^2$. Equality occurs in the latter case for **cap bodies** (Figure A14 illustrates such critical sets) obtainable as the convex hull of a sphere and certain exterior points. It is possible to have $y = 0$

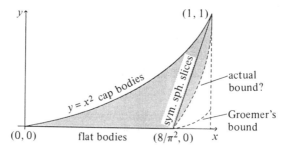

Figure A13.   The Blaschke diagram—the pairs (x, y) that can occur are shaded.

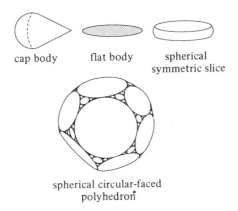

cap body   flat body   spherical
                                     symmetric slice

spherical circular-faced
polyhedron

Figure A14.   Critical convex bodies for Blaschke's problem.

if $0 \leq x \leq 8/\pi^2$; this is achieved by two-dimensional **flat bodies**. However, if $8/\pi^2 < x \leq 1$ then y must be strictly positive, and Groemer has proved that over this range

$$y \geq \frac{\pi}{8} x \left(x - \frac{8}{\pi^2} \right).$$

On the other hand, **symmetric spherical slices** give rise to points on the line segment joining $(8/\pi^2, 0)$ to $(1, 1)$. Moreover, Bieri has given examples with (x, y) strictly to the right of this line. Thus the outstanding problem is to find the minimum value of y for $8/\pi^2 < x \leq 1$. Do we get some parabola through $(8/\pi^2, 0)$ and $(1, 1)$, perhaps even the parabola that, when extended, passes through $(0, 0)$? Beiri suggests that the extremals are given by **spherical circular-face polyhedra** obtained by slicing off countably many disjoint circular caps from a sphere, to get a body whose surface is formed by a collection of contiguous disks. Good accounts of this problem are given in Hadwiger's book and in Sangwine-Yager's article.

W. Blaschke, Eine Frage über konvexe Körper, *Jahresber. Deutsch. Math. Verein* **25** (1916) 121–125.

H. Bieri, Über Konvexe Extremalkörper, *Experientia* **5** (1949) 355; *MR* **11**, 200.

H. Bieri, Mitteilung zum Problem eines konvexen Extremalkörpers, *Arch. Math. (Basel)* **1** (1949) 462–463; *MR* **11**, 127.

H. Groemer, Eine neue Ungleichung für konvexe Körper, *Math. Z.* **86** (1985) 361–364.

H. Hadwiger, P. Glur & H. Bieri, Die symmetrische Kugelzone als extremale Rotationskörper, *Experientia* **4** (1948) 304–305; *MR* **10**, 141.

H. Hadwiger, Elementare Studie über konvexe Rotationskörper, *Math. Nachr.* **2** (1949) 114–123; *MR* **11**, 127.

H. Hadwiger, [Had], Section 28.

H. Hadwiger, Notiz für fehlenden Ungleichung in der Theorie der konvexen Körper, *Elem. Math.* **3** (1948) 112–113; *MR* **10**, 395.

J. R. Sangwine-Yager, The missing boundary of the Blaschke diagram, *Amer. Math. Monthly* **96** (1989) 233–237.

A22. Minimal bodies of constant width.

A22. Minimal bodies of constant width. The d-dimensional set of constant width 1 of largest volume is the d-dimensional ball of radius $\frac{1}{2}$, and the plane set of constant width 1 of least area is the Reuleaux triangle. The minimum problem is unsolved in three or more dimensions. Even finding the 3-dimensional set of least volume presents formidable difficulties, not least because of the awkwardness of describing such sets. Chakerian proved that any set in $\mathbb{R}^3$ of constant width 1 has volume at least $\pi(3\sqrt{6} - 7)/3 = 0.365\ldots$. For many years it has been widely believed that Meissner's three-dimensional generalization of the Reuleaux triangle of volume $\pi(\frac{2}{3} - \frac{1}{4}\sqrt{3}\cos^{-1}(\frac{1}{3})) = 0.420\ldots$ is the unique extremal (see Bonnesen & Fenchel, page 135) though this has not been proved. Can these bounds for the minimum volume be improved?

Blaschke's relation for a set of constant width 1 with volume V and surface area S gives $S = 2V + \frac{2}{3}\pi$, so the problem is equivalent to that of minimizing the surface area.

The problem has been studied in higher dimensions by Firey, Chakerian, and Schramm. In particular, Schramm shows that the d-dimensional volume of any set of constant width one in $\mathbb{R}^d$ is at least $(\sqrt{3 + 2/(d + 1)} - 1)^d$ times the volume of the d-dimensional ball of diameter 1, though this figure is probably much too small for large d. Danzer has conjectured that the set of least volume of a given constant width in $\mathbb{R}^d$ always has the symmetry group of the regular d-simplex. In particular does the minimal 3-dimensional body of constant width have the symmetry of the regular tetrahedron?

T. Bonnesen & W. Fenchel, [BF].

G. D. Chakerian, Sets of constant width, *Pacific J. Math.* **19** (1966) 13–21; *MR* **34** #4986.

G. D. Chakerian & H. Groemer, Convex bodies of constant width, in [GW], 49–96; *MR* **85f**:52001.

W. J. Firey, Lower bounds for volumes of convex bodies, *Arch. Math. (Basel)* **16** (1965) 69–74; *MR* **31** #5152.

E. Meissner, Über Punktmengen konstanter Breite, *Vjsch. Naturforsch. Ges. Zürich* **56** (1911) 42–50.

E. Meissner, Drei Gipsmodelle von Flächen konstanter Breite, *Z. Math. Phys.* **60** (1912) 92–94.

O. Schramm, On the volume of sets having constant width, *Israel J. Math.* **63** (1988) 178–182.

A23. Constrained isoperimetric problems.

A23. Constrained isoperimetric problems. Let K_0 be a given convex region in the plane. Besicovitch considered the problem of finding the subset K of K_0 of given perimeter L for which the area $A(K)$ is maximum. He showed that the area is maximized by the (convex) set K_r for some value of r, where K_r consists of the union of all discs of radius r contained in K. Thus the extremal set is bounded by parts of the boundary of K_0 together with circular arcs all of the same radius. The 3-dimensional analog, of finding the subset of K_0 of largest volume with given surface area, is a challenging one. It is not difficult to see that the boundary of any extremal consists of parts of the boundary of

K_0 together with pieces of constant mean curvature, but rather more information is necessary to determine the extremals. In general they are *not* given by the union of all balls in K_0 of a particular radius—Herz & Kaapke showed this in the case where K_0 is a circular cylinder.

Given a plane convex region K_0, what is the convex subset K of K_0 for which the ratio $A(K)/L(K)$ is maximum? (We look for "large" subsets that are not too "elongated".) It follows from Besicovitch's work that the extremal is K_r for some r, and Singmaster & Souppouris showed that r is given as the unique number satisfying $A(K_r)/L(K_r) = r$. In three dimensions, how does one maximize $V(K)/S(K)$, given that K is contained in K_0? Are there any qualitative properties of the extremals that would help in solving the problem in particular cases?

Various authors have obtained explicit solutions to these problems in special cases, for example with K_0 as a regular polygon.

Of course there are many other problems of this type which might lead to interesting results. For example, we could try to maximize $A(K)/[L(K)]^\alpha$ for any fixed $0 < \alpha < 2$ with $K \subset K_0$. What about the constrained maximum of $A(K)$, given the width of K?

A. S. Besicovitch, Variants of a classical isoperimetric problem, *Quart. J. Math. Oxford* (2) **3** (1952) 42–49; *MR* **13**, 768.

H. T. Croft, Cushions, cigars and diamonds: an area-perimeter problem for symmetric ovals, *Math. Proc. Cambridge Philos. Soc.* **85** (1979) 1–16; *MR* **80e**:52001.

R. F. DeMar, A simple approach to isoperimetric problems in the plane, *Math. Mag.* **48** (1975) 1–12.

A. D. Garvin, A note on DeMar's "A simple approach to isoperimetric problems in the plane" and an epilogue, *Math. Mag.* **48** (1975) 219–221.

B. Herz & J. Kaapke, Ein isoperimetrisches Problem mit Nebenbedingung, *Elem. Math.* **28** (1973) 63–65.

T.-P. Lin, Maximum area under constraint, *Math. Mag.* **50** (1977) 32–34.

D. Singmaster & D. J. Souppouris, A constrained isoperimetric problem, *Math. Proc. Cambridge Philos. Soc.* **83** (1978) 73–82; *MR* **81i**:52011.

J. Steiner, Sur le maximum et le minimum des figures dans le plan sur la sphère et dans l'espaces en général, *J. Math. Pures Appl.* **6** (1841) 105–170.

A24. **Is a body fairly round if all its sections are?** The following set of problems, based on the maxim that "a convex body must be fairly spherical if its sections are fairly circular" seems to stem from Steinhaus. Is the following inequality true?

$$\frac{V^2}{S^3} \geq \frac{4\pi}{9} \inf_P \frac{A(K \cap P)^2}{L(K \cap P)^4},$$

where the infimum is taken over the planes P that intersect K in positive area? (Of course, the isoperimetric quotients A/L^2 and V^2/S^3 are measures of circularity and sphericity, respectively.) More generally, for which α is it true that

$$\frac{V^2}{S^3} \geq \frac{(4\pi)^\alpha}{36\pi} \inf_P \left(\frac{A(K \cap P)}{L(K \cap P)^2} \right)^\alpha \quad ?$$

It would be of interest to prove such inequalities even with smaller constants on the right-hand side.

Again, one could compare $S^3/V^2 - 36\pi$ with powers of $L^2/A - 4\pi$. Chakerian shows that

$$\frac{MV}{S^2} \geq \frac{\pi^3}{8} \inf_P \frac{A(K \cap P)}{L(K \cap P)^2},$$

where M is the integral mean curvature of K, and conjectures that $\frac{1}{8}\pi^3$ can be replaced by $\frac{4}{3}\pi$.

Do we get any results of interest if we consider only sections through the center of gravity?

In this connection it is perhaps appropriate to mention the remarkable theorem of Dvoretsky, that given $\varepsilon > 0$ and an integer k, there is an integer d ($d = \lfloor \exp(2^{15}\varepsilon^{-2}k^2 \ln k) \rfloor$ will do) such that every convex body in $\mathbb{R}^d$ has, through any given point, a k-dimensional section K_0 which is nearly spherical, in the sense that the ratio of the circumradius to the inradius of K_0 is at most $1 + \varepsilon$.

G. D. Chakerian, Is a body spherical if all its projections have the same I.Q.? *Amer. Math. Monthly* **77** (1970) 989–992.

A. Dvoretsky, Some near-sphericity results, in [Kle], 203–210; *MR* **28** #1533.

H. Groemer, On plane sections and projections of convex sets, *Canad. J. Math* **21** (1969) 1331–1337; *MR* **41** #2533.

H. Steinhaus, P232, *Colloq. Math.* **5** (1958) 264.

A25. How far apart can various centers be? Consider the centroid, the centroid of the perimeter, the circumcenter, and the incenter of a convex set. How far apart can any particular pair be in terms of the diameter, width, circumradius or inradius? Or (in two dimensions) in terms of the perimeter or the square root of the area?

The distance apart of the centroid and circumcenter in terms of the diameter was considered by Pál, by Grünbaum, and by Eggleston, culminating in a complete solution. Scott has found the maximum distance apart of the centroid and circumcenter in convex sets of a given circumradius.

This problem has obvious analogs in three or more dimensions.

H. G. Eggleston, An extremal problem concerning the centroid of a plane convex set, [Fen], 68.

H. G. Eggleston, The centroid and circumcentre of a plane convex set, *Compositio Math.* **21** (1969) 125–136; *MR* **39** #6173.

J. F. Pál, Om convexe figures tyndepunkter, *Mat. Tidsskrift* **B** (1937) 101–103.

P. R. Scott, Centre of gravity and circumcentre of a convex body in the plane, *Quart. J. Math. Oxford Ser.* (2) **40** (1989) 111–117; *MR* **89m**:52017.

A26. **Dividing up a piece of land by a short fence.** What is the shortest length of fencing required to divide a given piece of land into two parts of equal area? We may interpret the problem mathematically in several ways.

Let K be a plane convex set of area A. It is easy to prove that there is a chord of length at most $3^{1/4}A^{1/2}$ that bisects the area. Santaló asked us if there is always such a chord of length at most $(4/\pi)^{1/2}A^{1/2}$. Can one get more precise estimates by introducing w, D, r, or R? The corresponding problems for the maximal chord lengths have been discussed by Radizewski and Eggleston.

The problem becomes rather more intriguing if the "fence" is allowed to be curved. The shortest curve (necessarily a straight line or circular arc) that bisects the area of a unit-area centro-symmetric convex set K is longest if K is circular. Is this still true if the centro-symmetry condition is dropped? See Pólya.

Can one obtain good estimates for the length of fence required in terms of the usual measures of K? Various inequalities have been obtained by Hadwiger, by Iseli, and by Bokowski & Sperner. The best results to date are due to Bokowski, who shows, in the more general situation where K is divided into parts of areas A^- and A^+, that the length L of any dividing curve satisfies

$$D^3 L \ge A^- A^+$$

and also

$$DA^*L \ge A^- A^+,$$

where D is the diameter of K and A^* is the area of the difference set $\{\mathbf{x} - \mathbf{y} : \mathbf{x}, \mathbf{y} \in K\}$. The integral geometrical methods of Bokowski also extend to give inequalities in the higher-dimensional situation of a convex body divided into two parts of given d-dimensional volumes by a $(d-1)$-dimensional surface. Can any of these inequalities be refined?

For which plane sets K is the shortest dividing curve actually a straight line segment?

Santaló suggests another variant. Let K be a 3-dimensional body of surface area S. Is there always a curve on the surface of K of length at most $\sqrt{\pi S}$ that divides the surface area into equal parts, equality occurring for the surface of the sphere?

There are many variations on this theme. For example, what is the best way of dividing a piece of land into n equal parts by a configuration of fences, with total length as short as possible?

J. Bokowski, Ungleichungen für den Inhalt von Trennflächen, *Arch. Math. (Basel)* **34** (1980) 84–89; *MR* **81f**:52018.

J. Bokowski & E. Sperner, Jr., Zerlegung konvexer Körper durch minimale Trennflächen, *J. Reine Angew. Math.* **311/312** (1979) 80–100; *MR* **81b**:52010.

H. G. Eggleston, The maximal length of chords bisecting the area or perimeter length of plane convex sets, *J. London Math. Soc.* **36** (1961) 122–128; *MR* **23** #A2795.

L. M. Gysin, Inequalities for the product of the volumes of a partition determined in a convex body by a surface, *Rend. Circ. Mat. Palermo* (2) **35** (1986) 420–428; *MR* **89f**:52029.

H. Hadwiger, Über die Flächeninhalte ebener Schnitte konvexer Körper, *Elem. Math.*
30 (1975) 97–102; *M R* **52** #6578.

M. Iseli, Über die Flächeninhalte ebener Schnitte konvexer Körper, *Elem. Math.* **33**
(1978) 129 –134; *M R* **80c**:52010.

G. Pólya, Aufgabe 283, *Elem. Math.* **13** (1958) 40–41.

K. Radziszewski, Sur les cordes qui partagent l'aire d'un ovale en 2 parties égales, *Ann.
Univ. Mariae Curie-Sklodowska Sect. A* **8** (1954) 89–92; *M R* **18**, 330.

K. Radziszewski, Sur les cordes qui partagent le périmètre d'un ovale en 2 parties
égales, *Ann. Univ. Mariae Curie-Sklodowska Sect. A* **8** (1954) 93–96; *M R* **18**, 330.

L. A. Santaló, An inequality between the parts into which a convex body is divided by
a plane section, *Rend. Circ. Mat. Palermo* (2) **32** (1983) 124–130; *M R* **85c**:52020.

A27. Midpoints of diameters of sets of constant width.

Let K be a d-dimensional set of constant width 1. Let M be the collection of midpoints of the diametral chords of K, that is, the chords of K of length 1. What is the largest possible value of the diameter of M, taken over all such K? Can one always find a d-dimensional simplex containing M and contained in K? This last question is likely to be hard since an affirmative answer would imply the truth of Borsuk's conjecture (see Section D14).

Z. A. Melzak, Problems connected with convexity, Problem 19, *Canad. Math. Bull.* **8**
(1965) 565–573; *M R* **33** #4781.

Z. A. Melzak, More problems connected with convexity, Problem 7, *Canad. Math. Bull.*
11 (1968) 482–494; *M R* **38** #3767.

A28. Largest convex hull of an arc of given length.

An exercise using the isoperimetric inequality shows that the plane arc of given length whose convex hull has greatest area is of semicircular shape. (First observe that the maximal area for any given position of the end-points results from a circular arc, and then maximize this as the distance between the end points varies. Moran gives an alternative proof.) Into what shape should a piece of wire of length 1 be bent in order to maximize the volume of its (three-dimensional) convex hull? Egerváry has solved this under the assumption that the curve contains no four coplanar points; the extremal is then one turn of a circular helix of height $1/\sqrt{3}$ and base radius $1/\pi\sqrt{6}$. Is this the curve of largest convex hull in the general case?

Bonnesen & Fenchel ask how to bend a *closed* loop of wire so as to maximize the volume of the convex hull. Under restrictive symmetry conditions, this problem is treated by Melzak and also by Schoenberg (in even-dimensional space). The question of maximizing the surface area of the convex hull does not seem to have been considered. Nor does the problem of maximizing the area or volume of the convex hull of some connected arrangement of pieces of wire of total length one. Is this always achieved by an arc?

Adhikari & Pitman show that a planar arc of length one has greatest minimal width when it is in the shape of a "caliper." They ask whether the set formed by the three arcs joining the center of an equilateral triangle to its vertices has greatest minimal width among all connected sets of given length. They also raise analogous questions in three dimensions: "What is the shortest length of wire that can be bent so as not to be able to fall into a parallel-sided space between your cooker and the kitchen wall?"

A. Adhikari & J. Pitman, The shortest planar arc of width 1, *Amer. Math. Monthly* **96** (1989) 309–327.

T. Bonnesen & W. Fenchel, [BonFe], 111.

E. Egerváry, On the smallest convex cover of a simple arc of space-curve, *Publ. Math. Debrecen* **1** (1949) 65–70; *MR* **12**, 46.

Z. A. Melzak, The isoperimetric problem of the convex hull of a closed space-curve, *Proc. Amer. Math. Soc.* **11** (1960) 265–274; *MR* **22** #7058.

Z. A. Melzak, Numerical evaluation of an isoperimetric constant, *Math. Comput.* **22** (1968) 188–190; *MR* **36** #7023.

P. A. P. Moran, On a problem of S. Ulam, *J. London Math. Soc.* **21** (1946) 175–179; *MR* **8**, 597.

I. J. Schoenberg, An isoperimetric inequality for closed curves in even-dimensional Euclidean spaces, *Acta Math.* **91** (1954) 143–164; *MR* **16**, 508.

A29. Roads on planets.

It is required to build a road on a newly conquered planet in such a way that no point on the planet is further than distance ε from the road. More exactly, what is the shortest curve on the surface of a unit sphere that comes within distance ε of every point on the surface. Is it always the "apple-peeling" spiral? Can one do better if a network consisting of connected pieces of curve is used?

The same question may be asked for the shortest paths passing within distance ε of all points on more general surfaces or in plane domains. Is it true that the length of the shortest curve on a convex surface equals $\frac{1}{2}A\varepsilon^{-1} + o(\varepsilon^{-1})$, where A is the surface area, and can one get a better expression for the error in terms of the usual measures of the surface? What properties do such minimal paths share?

A related problem appears in Melzak's collection, where it is required to find the shortest path whose convex hull approximates a convex surface to within ε.

Z. A. Melzak, Problems connected with convexity, Problem 39, *Canad. Math. Bull.* **8** (1965) 565–573; *MR* **33** #4781.

A30. The shortest curve cutting all the lines through a disk.

Croft considered the problem of finding the shortest curve in the plane that cuts every chord (produced both ways) of a fixed disk K of unit radius. The nature of the problem depends in which of the following ways curve is interpreted: (a) a continuous (rectifiable) curve; (b) a countable collection of rectifiable curves somehow connected to each other; and (c) a countable collection of rectifiable curves not necessarily connected to each other.

Compactness arguments show that extremals exist in cases (a) and (b). Croft conjectured that in both cases the (essentially) unique extremal is of length $2 + \pi = 5.14159\ldots$ [Figure A15(a)]. This was proved in case (a) by Joris and by Faber, Mycielski & Pedersen and in case (b) by Eggleston. A short proof in either case would be welcome—we feel that there ought to be a snappy integral geometric argument. Faber, Mycielski & Pedersen give a two-piece

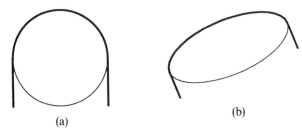

Figure A15.   The shortest curves cutting every line through (a) a circle and (b) an ellipse.

curve of length 4.819... for (c) but it seems possible that we can always reduce the total length by increasing the number of pieces. It might nevertheless be possible to obtain an expression for the infimum length.

Of course, we can ask the same question for any convex set K. Faber & Mycielski give short curves in cases (a), (b), and (c) for various K; in particular for regular n-gons for small n. For many K, but by no means all, the shortest continuous curve (a) is given by a portion of the boundary curve of K extended by a line segment at each end to meet some tangent of K perpendicularly, as in Figure A15(b). It would be interesting to know for which K the shortest curve is of this form; perhaps this is so if the diameter/width ratio of K is not too large.

Faber & Mycielski point out that when K is a circular disk the problem is equivalent to that of a telephone company who wish to dig a trench to locate a straight cable that is known to pass within a yard, say, of a given marker.

A closely connected set of "practical" problems is due to S. Burr (see Ogilvy). A swimmer is either (1) in the center of a parallel sided river of known width, or (2) at sea a known distance from a straight shoreline, when fog descends, so that the direction of the shore(s) is unknown. Assuming that he can navigate correctly, what path should be followed to find land so as either (i) to minimize the longest time to shore, or (ii) to minimize the expected time taken (assuming all directions of the shore equally likely), or (iii) to maximize the chance of reaching the shore given that he can only swim a certain distance?

More formally, given a unit disk, we seek continuous curves with one end at the center cutting some or all of its tangents [where in (2) all tangents are on an equal footing and in (1) we are only concerned with one—either one—of the two parallel tangents in each direction], and satisfying some extremal property. It is conjectured that for problems (1i) and (2i) and possibly also (1ii) and (2ii) the extremal curves are as shown in Figure A16. Of course in case (iii) the curve will depend on the maximum distance that can be swum.

Bellman has suggested a generalization of the swimmer problem; that of a man lost in a forest F. There is a plethora of problems depending not only on the shape of F and his position in it, but also on his knowledge of the shape and orientation and his initial position. Some versions are equivalent to finding the infimum length of a path, no congruent copy of which can be

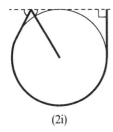   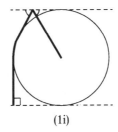

(2i)   (1i)

Figure A16.   Conjectured shortest paths for problems (2i) and (1i).

covered by F, and so are closely connected with the "worm problems" of Section D18.

Of course, the original problem has higher dimensional variants. Faber & Mycielski give a surface of area $2\pi + \frac{1}{2}\pi^2 + \varepsilon$ that cuts every line cutting the unit sphere. Can this value be reduced? Perhaps more interestingly, what is the shortest continuous curve which intersects each plane cutting the sphere, that is the shortest curve with convex hull containing the sphere (compare with Section A28)? Is it half a revolution of a helix with a straight line segment at each end?

For many further problems of this type see Croft and Faber & Mycielski.

H. T. Croft, Curves intersecting certain sets of great circles on the sphere, *J. London Math. Soc.* (2) **1** (1969) 461–469; *MR* **40** #865.

H. G. Eggleston, The maximal inradius of the convex cover of a plane connected set of given length, *Proc. London Math. Soc.* (3) **45** (1982) 456–478; *MR* **84e**:52004.

V. Faber & J. Mycielski, The shortest curve that meets all the lines that meet a convex body, *Amer. Math. Monthly* **93** (1986) 796–801; *MR* **87m**:52017.

V. Faber, J. Mycielski & P. Pedersen, On the shortest curve which meets all the lines which meet a circle, *Ann. Polon. Math.* **44** (1984) 249–266; *MR* **87b**:52023.

H. Joris, Le chasseur perdu dans la forêt, *Elem. Math.* **35** (1980) 1–14; *MR* **81d**:52001.

C. S. Ogilvy, *Tomorrow's Math*, Oxford University Press, Oxford, 1962, 23–24.

R. Schnieder & J. A. Wieacker, Einschliessung ebener Kurven, *Elem. Math.* **40** (1985) 98–99; *MR* **86m**:52014.

A31.   Cones based on convex sets. Let K be a convex set contained in a plane P in 3-dimensional space. Given $h > 0$ and a point $\mathbf{x}$ in K, form the cone based on K with vertex distance h up the perpendicular to P through $\mathbf{x}$, and let $A(\mathbf{x}, h)$ denote the surface area of this cone (as in Figure A17). For fixed h there is a unique $\mathbf{x}(h)$ for which $A(\mathbf{x}, h)$ attains a minimum. (It is easy to see that if there were minima at two distinct points, the area would take a strictly smaller value at their midpoint.) How are the points $\mathbf{x}(h)$ related to K? If K is any polygon circumscribing a circle, then $\mathbf{x}(h)$ is the center of the circle. Is there any simple description of these points for other plane convex sets? In particular, is there a simple description of $\lim_{h \to 0} \mathbf{x}(h)$ and $\lim_{h \to \infty} \mathbf{x}(h)$ in terms of the geometry of K?

A. Pleijel, Problem 16, *Math. Scand.* **3** (1955) 306.

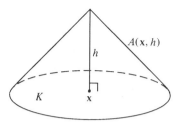

Figure A17.   For which x is the surface area $A(x, h)$ minimum?

A32.   Generalized ellipses.

A standard method of drawing an ellipse is to move a pen around inside a taut loop of string round pins located at the foci. This reflects the fact that the equation of an ellipse may be written $|x - x_1| + |x - x_2| = c$, where x_1, x_2 are the foci and c is a constant. A curve C formed by those x satisfying $\sum_{i=1}^{n} |x - x_i| = c$, where $x_1, \ldots, x_n$ are fixed points in the plane is called a **multifocal ellipse**. C is a convex curve provided that $c > c_0$, where c_0 is the minimum of $\sum_{i=1}^{n} |x - x_i|$ (attained at a unique point in the plane). Many years ago Weissfeld asked if every closed convex curve could be approximated arbitrarily closely by multifocal ellipses if x_i and c are suitably chosen; the problem was repeated by Hammer. Erdős & Vincze, and also Danzer showed that the equilateral triangle was not approximable by multi-focal ellipses. Subsequently, Erdős & Vincze showed that it is not possible to approximate any convex curve containing more than one line segment, though there are approximable curves containing just one line segment. Exactly which curves can be approximated by multifocal ellipses? Erdős & Vincze point out that any curve of the form

$$\int_E |x - y| \, dA(y) + \int_\Gamma |x - y| \, dl(y) + \sum_i |x - x_i| = c, \qquad (c > c_0)$$

can be approximated if E is some open plane region, and Γ is a rectifiable, plane curve. Does this describe all the approximable curves?

Is the situation radically different if we consider approximation by **multi-focal hyperbolas** of the form $\sum_{i=1}^{n} |x - x_i| - \sum_{i=1}^{m} |x - y_i| = c$?

In three dimensions, which surfaces can be approximated by **multifocal ellipsoids** with equations $\sum_{i=1}^{n} |x - x_j| = c$?

Hammer has suggested problems of the following type: Let C be the curve traced by holding a pencil against a taut loop of string around a convex set K (thus if C is a line segment we get an ellipse). How are the properties of C (width, area, length, diameter, etc.) related to those of K? Are there any non-trivial inequalities to be discovered? What conditions on C guarantee that K is a disk or set of constant width? If the length of the loop is large compared with the diameter of K then C must be "fairly circular." Can this be formulated more precisely?

Similar questions can be asked if C consists of the points x such that the area enclosed by the tangents to K from x is constant.

L. Danzer, To Problem 17, [Fen], 318.

P. Erdős & I. Vincze, On the approximation of convex, closed plane curves, *Mat. Lapok.* **9** (1958) 19–36; *MR* **20** #6070.

P. Erdős & I. Vincze, On the approximation of convex, closed plane curves by multifocal ellipses, *J. Appl. Prob.* (Special Vol.) **19** (1982) 89–96; *MR* **83a**:52006.

P. C. Hammer, Problem 13, [Fen], 316.

P. C. Hammer, Problem 17, [Fen], 318.

P. Hartman & F. A. Valentine, On generalized ellipses, *Duke Math. J.* **26** (1959) 373–385; *MR* **21** #6566.

E. Weissfield, Sur le point pour lequel la somme des distances de n points donnés est minimum, *Tôhoku Math. J.* **43** (1937) 355–386.

T. Zamfirescu, Elipsoïdes et hyperboloïdes géneralisés, *Atti. Accad. Sci. Torino Cl. Sci. Fis. Mat. Natur.* **118** (1984) 314–324; *MR* **89b**:52005.

A33. Conic sections through five points.

In this section we group together several problems that depend on the fact that five points in the plane in general position determine a unique conic section.

Given six points in the plane not all lying on a single conic, can the six conics through each subset of five of the points all be similar to each other? They cannot all be congruent—the various possibilities have been eliminated by Kelly (equilateral hyperbola), Seidel & von Vollenhoven (ellipse), Blumenthal (pair of lines), Pedoe (parabola), and Olson (arbitrary hyperbola).

Reznick wants to know what can be said about an infinite set S, in particular a union of rectifiable curves, if the conic defined by every five-point subset of S is an ellipse (or, respectively, a hyperbola). It is not hard to show that if there is some line that cuts S in three points near each of which S is a continuous curve, then both ellipses and hyperbolas will occur, and this suggests that the set sought will be formed of pieces of convex curve. It is suggested that $|x|^{2.001} + |y|^{2.001} = 1$ yields only ellipses.

L. M. Blumenthal, *Theory and Applications of Distance Geometry*, 2nd ed., Chelsea, New York, 1970, 153; *MR* **42** #3678.

L. M. Kelly, Covering problems, *Math. Mag.* **19** (1944) 123–130; *MR* **6**, 183.

R. C. Olson, On the six conics problem, *Discrete Math.* **10** (1974) 279–300; *MR* **50** #3092.

R. C. Olson, Solutions of elementary problems, *Amer. Math. Monthly* **83** (1976) 285.

D. Pedoe, Thinking geometrically, *Amer. Math. Monthly* **77** (1970) 711–721.

J. J. Schäffer, Ungelöste Problem 43, *Elem. Math.* **17** (1962) 65.

J. J. Seidel & J. van Vollenhoven, Zum Probleme 43, *Elem. Math.* **17** (1962) 85.

A34. The shape of worn stones.

In an entertaining paper Firey gives a mathematical formulation for the wearing of stones on a beach by abrasion against each other. He shows, under quite plausible assumptions, that a stone which is initially convex and centro-symmetric converges to a spherical shape. He suggests that it might be possible to modify the results to apply to stones that are not initially centro-symmetric. Rogers conjectures that the stones should approach an ellipsoidal shape more rapidly than they approach a spherical shape.

W. J. Firey, Shapes of worn stones, *Mathematika* **21** (1974) 1–11; *MR* **50** #14487.

C. A. Rogers, Problem 4, *Bull. London Math. Soc.* **8** (1976) 29.

A35. Geodesics.

We take, at least initially, a sufficiently smooth convex surface C in $\mathbb{R}^3$. For our purpose, we regard a **geodesic arc** as any curve in C of shortest length between two points on the surface.

(i) Is the sphere the only surface for which, whenever points can be joined by two distinct geodesics, then they can be joined by an infinity of geodesics? This problem is due to Gotz & Rybarski.

(ii) Steinhaus asked whether the sphere is characterized by the property that the mapping associating points on C with their most distant points (via the shortest geodesic) is 1–1 and symmetrical.

(iii) Steinhaus showed that there always exist at least two distinct paths from a point $\mathbf{x}$ of C to its farthest point (supposed unique). Is there always some $\mathbf{x}$ on C with three such paths?

(iv) Steinhaus also asked us what can be said qualitatively about the set of all "farthest points" from a point $\mathbf{x}$. Examples, such as taking $\mathbf{x}$ as a middle point on a long circular cylinder, show that it need not be connected. Must this set be a single point for "almost all" $\mathbf{x}$ on C?

(v) If all the geodesics from each point of C meet in a second point, must C be spherical?—See Hilbert & Cohn-Vossen. It is known that this is so (indeed in the more general setting of a Riemann manifold) if the geodesic distances to the second meeting points are constant. Blaschke termed such surfaces "Wiedersehensflächen" (a group of people setting out from a point along different geodesics paths at the same speed all "meet again") and conjectured that such surfaces were spherical. This was proved by Green.

(vi) The theorem of Lyusternik & Schnirelmann states that any surface has at least three non-self-intersecting closed geodesics. Burton showed for any (not necessarily smooth) convex surface C that if every geodesic lies in a plane then C must be spherical and Takeuchi showed that, for a C^∞ surface, this conclusion follows if there are four planar geodesics through each point.

(vii) Armstrong asked if the sphere is the only (sufficiently smooth) body of constant width all of whose geodesics are closed.

Many of these questions can be asked for arbitrary (non-smooth) convex surfaces, or for non-convex surfaces homeomorphic to a sphere. The most natural setting for this type of problem is on a Riemannian manifold (see Busemann or Klingenburg).

W. W. Armstrong, Curves on surfaces of constant width, *Canad. Math. Bull.* **9** (1966) 15–22; *MR* **33** #6511.

W. Blaschke, *Einführung in die Differentialgeometrie*, Springer-Verlag, Berlin, 1950; *MR* **13**, 274.

G. R. Burton, Convex surfaces whose geodesics are planar, *Monatsh. Math.* **84** (1977) 177–195; *MR* **58** #7506.

H. Busemann, *The Geometry of Geodesics*, Academic Press, New York, 1955; *MR* **17**, 779.

A. Götz & A. Rybarski, Problem 102, *Colloq. Math.* **2** (1951) 301–302.

L. Green, Auf Wiedersehensflächen, *Ann. Math.* (2) **78** (1963) 289–299; *MR* **27** #5206.

D. Hilbert & S. Cohn-Vossen, *Geometry and the Imagination*, Chelsea, New York, 1952, 224.

W. Klingenberg, *Lectures on Closed Geodesics*, Springer-Verlag, Berlin, 1978; *MR* **57** #17563.

L. Lyusternik & L. Schnirelmann, Sur la problème de trois géodésiques fermées sur les surfaces de genre 0, *C. R. Acad. Sci. Paris* **189** (1929) 269–271.

H. Steinhaus, Problem 291, *Colloq. Math.* **7** (1959) 110.

H. Steinhaus, On shortest paths on closed surfaces, *Bull. Acad. Polon. Sci. Sér. Sci. Math. Astr. Phys.* **6** (1958) 303–308; *MR* **20** #3255.

N. Takeuchi, A surface which contains planar geodesics, *Geom. Dedicata* **27** (1988) 223–225.

A36.   Convex sets with universal sections. Does there exist a compact convex body K in $\mathbb{R}^3$ such that every (closed) plane convex subset of the unit disk is congruent to some plane section of K? This question, due to Ryll–Nardzewski, is not as far-fetched as it at first seems, since there exists a 4-dimensional K with this property. In fact, Grząślewicz uses a Peano curve in the metric space of convex sets to construct a d-dimensional convex set with a section congruent to every sufficiently small $(d - 2)$-dimensional convex set. Does there at least exist a convex body in $\mathbb{R}^3$ with plane sections similar to, or even just affinely equivalent to, all plane convex sets? Does the situation change if K is allowed to be unbounded?

Grünbaum showed that no centro-symmetric body K in $\mathbb{R}^3$ can have a central plane section affinely equivalent to each (proper) centro-symmetric plane convex set, answering a Scottish Book question of Mazur. Further, Bessaga showed that this was not possible for any finite-dimensional centro-symmetric K.

Among many possible other variants, Melzak has constructed a convex **pseudopolyhedron** (that is, the convex hull of a convergent sequence of points in $\mathbb{R}^3$) that has a plane section similar to every triangle, but Shephard has shown that it is not possible to have a section similar to every polygon. Klee has considered the problem of finding convex polyhedra with sections affinely equivalent to certain families of polygons.

Does there exist, for some d, a convex subset of $\mathbb{R}^d$ with universal projections, (i.e., with some plane projection similar to, or even congruent to, each plane convex set)?

C. Bessaga, A note on universal Banach spaces of finite dimension, *Bull. Acad. Polon. Sci. Sér. Sci. Math. Astr. Phys.* **6** (1958) 97–101; *MR* **22** #4935.

B. Grünbaum, On a problem of S. Mazur, *Bull. Res. Council Israel* (Section F) **7** (1957/58) 133–135; *MR* **21** #4347.

R. Grząślewicz, A universal convex set in Euclidean space, *Colloq. Math.* **45** (1981) 41–44; *MR* **83m**:52009.

V. Klee, Polyhedral sections of convex bodies, *Acta. Math.* **103** (1960) 243–267; *MR* **25** #2512.

J. Lindenstrauss, Note on Klee's paper "Polyhedral sections of convex bodies," *Israel J. Math.* **4** (1966) 235–242; *MR* **35** #703.

R. D. Mauldin, Problem 41, [Mau].
Z. A. Melzak, Limit sections and universal points of convex surfaces, *Proc. Amer. Math. Soc.* **9** (1958) 729–734; *MR* **20** #4237.
Z. A. Melzak, A property of convex pseudopolyhedra, *Canad. Math. Bull.* **2** (1959) 31–32; *MR* **21** #320.
G. C. Shephard, On a conjecture of Melzak, *Canad. Math. Bull.* **7** (1964) 561–563; *MR* **30** #505.

A37. Convex space-filling curves.

In 1890 Peano published his construction for a "space-filling curve," (that is a continuous mapping from the interval $[0, 1]$ onto the unit square $[0, 1] \times [0, 1]$). Mihalik and Wieczorek asked whether such curves could be found so that all subcurves formed convex sets. In other words, we seek a continuous function $f: [0, 1] \to \mathbb{R}^2$ such that for every $0 \le a < b \le 1$ the image under f of the interval $[a, b]$ is a plane convex set of positive area.

Pach & Rogers have constructed a function that maps the intervals $[0, b]$ and $[a, 1]$ onto convex sets for all a and b but the more general problem seems awkward.

J. Pach & C. A. Rogers, Partly convex Peano curves, *Bull. London Math. Soc.* **15** (1983) 321–328; *MR* **84m**:54036.
G. Peano, Sur une courbe qui remplit une aire plane, *Math. Ann.* **36** (1890) 157–160.

A38. m-convex sets.

A subset E of $\mathbb{R}^d$ is called m-**convex** if, for any m points $x_1, \ldots, x_m$ of E, there is some pair for which the line segment $[x_i, x_j]$ lies entirely in E. (A 2-convex set is just a convex set; a 3-convex set is sometimes called **Valentine convex**.)

Valentine and Eggleston showed that a plane 3-convex set is a union of at most three convex sets. Eggleston proved that for each m, a compact m-convex set in the plane is the union of a finite number of convex sets, and Breen & Kay established the quantitative estimate that a closed plane m-convex set is the union of a most $(m - 1)^3 3^{m-3}$ convex sets. Can this bound be improved? Examples show that it must be at least $\frac{1}{4} m^{3/2}$. Breen has similar estimates for non-closed sets.

On the other hand, Eggleston constructed a 4-dimensional 3-convex set not expressible as the union of finitely many convex sets (though such sets must have a countable decomposition into convex parts). The situation in three dimensions is unclear. Is every 3-dimensional 3-convex set a finite union of convex sets?—if so, of how many?

The interior of a convex set is clearly convex, and it follows from the result of Valentine and Eggleston above that the interior of a plane 3-convex set is 3-convex. On the other hand, for $m \ge 4$ there are m-convex sets with interiors that are just $(m + 1)$-convex. Tattersall asked if there was a bound $f(m)$ such that the interior of an m-convex set is always $f(m)$-convex. Leader has produced a 6-convex set with interior not m-convex for any m at all. Does there exist a 5-convex set, or even a 4-convex set whose interior fails to be m-convex for every m?

Calvert showed that if E and F are plane compact 3-convex sets then their intersection $E \cap F$ is 5-convex and Breen obtained conditions for intersections to be 3-convex. Eggleston showed that the vector sum $E + F$ $(= \{x + y : x \in E, y \in F\})$ is also 5-convex. Again, these results fail in four dimensions, with the 3-dimensional case unresolved. Can anything be said about intersections and sums of m-convex sets for larger m?

This idea may be generalized by calling a set E (m, n)-**convex** $(1 \leq n \leq \binom{m}{2})$ if, whenever $\mathbf{x}_1, \ldots, \mathbf{x}_m$ are m points of E, at least n of the $\binom{m}{2}$ segments $[\mathbf{x}_i, \mathbf{x}_j]$ are contained in E. However, Kay & Guay point out that this generalization is illusory, since, by a graph-theoretical result of Turán, any (m, n)-convex set is k-convex for a value of k that can be computed from m and n.

Of course, it would be possible to set up a theory of m-convex sets as has been done for convex sets. For example, what are the analogs of Helly's theorem (see Section E1) for such sets?

M. Breen, Decompositions for nonclosed planar m-convex sets, *Pacific J. Math.* **69** (1977) 317–324; *M R* **57** #10597.

M. Breen, Decomposition theorems for 3-convex subsets of the plane, *Pacific J. Math.* **53** (1974) 43–57; *M R* **50** #5628.

M. Breen, Intersectional properties concerning planar m-convex sets, *Israel J. Math.* **36** (1980) 83–88; *M R* **81h**:52006.

M. Breen, LNC points for m-convex sets, *Int. J. Math. Math. Sci.* **4** (1981) 513–528; *M R* **84**:52008.

M. Breen & D. Kay, General description theorems for m-convex sets in the plane, *Israel J. Math.* **8** (1970) 39–52.

D. I. Calvert, Generalizations of convexity, Thesis, University of London, 1979.

H. G. Eggleston, A condition for a compact plane set to be a union of finitely many convex sets, *Proc. Cambridge Philos. Soc.* **76** (1974) 61–66; *M R* **49** #7919.

H. G. Eggleston, Valentine convexity in n dimensions, *Math. Proc. Cambridge Philos. Soc.* **80** (1976) 223–228; *M R* **53** #14310.

H. G. Eggleston, A proof of a theorem of Valentine, *Math. Proc. Cambridge Philos. Soc.* **77** (1975) 525–528; *M R* **50** #14483.

H. G. Eggleston, Vector sums of Valentine convex sets, *Math. Proc. Cambridge Philos. Soc.* **92** (1982) 17–19; *M R* **84j**:52001.

D. C. Kay & M. D. Guay, Convexity and a certain property P_m, *Israel J. Math.* **8** (1970) 39–52; *M R* **46** #2385.

J. J. Tattersall, A Helly order for a family of (m, n)-convex sets, *Ars Combinatorica A***20** (1985) 153–162; *M R* **87e**:52015.

F. A. Valentine, A three point convexity property, *Pacific J. Math.* **7** (1957) 1227–1235; *M R* **20** #6071.

B. Polygons, Polyhedra, and Polytopes

In this chapter, we specialize from general convex sets to convex polygons and polyhedra. Their additional structure leads to many further properties worthy of particular study. Few can fail to appreciate the elegance and symmetry of polyhedral models. At the other extreme, deep and sophisticated mathematical techniques have been developed to explore their geometrical and combinatorial structure.

A 2-dimensional set that is the convex hull of a finite set of points is called a **convex polygon** [see Figure B1(a)]. A boundary point **x** is a **vertex** if there is a line that intersects the polygon in the single point **x**; thus a vertex is a boundary point where the internal angle is less than 180°. The line segments joining consecutive vertices are termed **edges**. Equivalently, a convex polygon is a convex subset of the plane bounded by a finite number of straight line segments. We often refer to a polygon with n vertices (and n sides) as an n-**gon**; it is **regular** if the edges are of equal length and the internal angles are all equal to $\pi(1 - 2/n)$.

We call a 3-dimensional convex hull of a finite set of points a **convex polyhedron** [see Figure B1(b)]. A boundary point **x** is a **vertex** if there is some plane that intersects the polyhedron P in the single point **x**. A line segment L in the boundary is an **edge** of P if there is a plane that intersects P in the segment L, and a region in the boundary P is a **face** if it is the intersection of a plane with P and has positive area. Euler's famous relation states that if a polyhedron has v vertices, e edges, and f faces then $v - e + f = 2$. The books by Coxeter, by Cundy & Rollett, and by Wenninger contain many pictures of polyhedra.

A polyhedron is said to be **simple** if exactly three edges terminate at every vertex. More generally, a polyhedron is k-**valent** if k edges are incident to each vertex. A polyhedron is **simplicial** if all its faces are triangles.

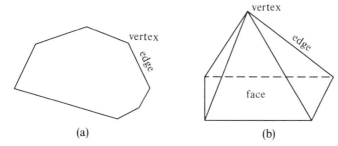

Figure B1.   (a) A convex polygon and (b) a convex polyhedron.

Occasionally we will want to refer to the higher-dimensional analogs. The d-dimensional convex hull of a finite set of points is termed a d-**polytope** (or just a **polytope** if the dimension is clear). We define k-**dimensional faces** $(0 \le k \le d - 1)$ in terms of the intersection of the boundary with $(d - 1)$-dimensional planes or hyperplanes. Thus the 0-faces are the vertices, the 1-faces are edges, etc; the $(d - 1)$-faces are sometimes called **facets**.

A d-polytope is a d-**simplex** if it is the convex hull of just $d + 1$ points not all in the same hyperplane. Thus a 2-simplex is a **triangle**, and a 3-simplex is a **tetrahedron**.

The words "polyhedron" and "polytope" are not always used consistently among authors. We shall use "polyhedron" for 3-dimensional bodies, while reserving "polytope" for higher dimensions. All our polygons, polyhedra, and polytopes are bounded.

The subject of non-convex polyhedra is much more complex than the convex case. There are difficulties even with the definitions of the basic terms, (e.g., edges, faces, etc). While it is possible to give analogs of many of our problems for non-convex polyhedra, we usually restrict our attention to the convex case. Thus, when the context is clear, we often write "polyhedron" rather than "convex polyhedron."

We shall sometimes be interested in the "combinatorial structure" of polyhedra. Two convex polyhedra are **combinatorially equivalent** or of the same **combinatorial class** if they have the same relative arrangement of vertices, faces, and edges. Thus, for example, any two tetrahedra are equivalent, and the cube is equivalent to any parallelepiped. More exactly, P and Q are equivalent if there is a mapping φ from the set S_P of vertices, edges, and faces of P to the set S_Q of vertices, edges, and faces of Q such that $\varphi(F_1) \subseteq \varphi(F_2)$, if and only if $F_1 \subseteq F_2$. (Inclusion here is in the set theoretic sense. Thus an edge "is contained in" a face, if and only if the edge is in the boundary of that face.) A similar definition holds for the equivalence of d-polytopes.

Steinitz proved what is sometimes referred to as the "fundamental theorem of polyhedra", namely, that a graph (i.e., a collection of points some pairs of which are joined by edges) G is realizable by the vertices and edges of a convex polyhedron, if and only if G is planar and 3-connected (i.e., at least three edges need to be cut to split the graph into two pieces). Thus many of the combinatorial questions concerning convex polyhedra may be interpreted as problems on planar 3-connected graphs.

Some remarkable progress has been made on the combinatorial structure of polytopes in recent years, for example with the proof of the upper bound conjecture and the characterization of f-vectors for simplicial polytopes. Much of this work is highly technical; the interested reader will find relevant information in the references that follow. In this book we have tried to list problems which have intuitive appeal.

Of the references listed, the book by Grünbaum remains the most comprehensive mathematical introduction to convex polyhedra. Many of the books and conference proceedings listed in the introduction to the convexity chapter also contain material on polyhedra.

A. D. Aleksandrov, *Konvexe Polyeder*, Akademie Verlag, Berlin, 1958; *MR* **19**, 1192.

A. Brønsted, *An Introduction to Convex Polytopes*, Springer-Verlag, New York, 1983; *MR* **84d**:52009.

H. S. M. Coxeter, *Regular Polytopes*, 3rd. ed., Dover, New York, 1973; *MR* **51** #6554.

H. S. M. Coxeter, *Regular Complex Polytopes*, Cambridge University Press, Cambridge, 1974; *MR* **51** #6555.

H. S. M. Coxeter, P. DuVal, H. T. Flather & J. F. Petrie, *The Fifty-nine Icosahedra*, Springer, Berlin, 1982; *MR* **84a**:52010.

H. M. Cundy & A. P. Rollett, *Mathematical Models*, 2nd ed., Clarendon, Oxford, 1961; *MR* **23** #A1484.

L. Fejes-Tóth, [Fej].

L. Fejes-Tóth, [Fej'].

B. Grünbaum, [Gru].

B. Grünbaum, Polytopes, graphs and complexes, *Bull. Amer. Math. Soc.* **76** (1970) 1131–1201; *MR* **42** #959.

B. Grünbaum & G. G. Shephard, Convex polytopes, *Bull. London Math. Soc.* **1** (1969) 257–300; *MR* **40** #3428.

H. Hadwiger, *Vorlesungen über Inhalt, Oberfläche und Isoperimetrie*, Springer, Berlin, 1957; *MR* **21** #1561.

L. A. Lyusternik, *Convex Figures and Polyhedra*, Dover, New York, 1963; Heath, Boston, 1966; *MR* **19**, 57; **28** #4427; **36** #4435.

P. McMullen & G. C. Shephard, *Convex Polytopes and the Upper Bound Conjecture*, Cambridge University Press, Cambridge, 1971; *MR* **46** #791.

S. A. Robertson, *Polytopes and Symmetry*, London Math. Soc. Lecture Notes 90, Cambridge University Press, Cambridge, 1984; *MR* **87h**:52012.

M. Senechal & G. Fleck (eds.), *Shaping Space*, Birkhäuser, Boston, 1988.

G. C. Shephard, Twenty problems on convex polyhedra, I, II, *Math. Gaz.* **52** (1968) 136–147, 359–367; *MR* **37** #6833, **38** #5111.

E. Steinitz, *Vorlesungen über die Theorie der Polyeder*, Springer, Berlin, 1934.

M. J. Wenninger, *Polyhedron Models*, Cambridge University Press, Cambridge, 1971; *MR* **57** #7350a.

B1. Fitting one triangle inside another.

Let triangles P and Q have edge lengths p_1, p_2, p_3 and q_1, q_2, q_3. Steinhaus asks for necessary and sufficient conditions on the p_i and q_i for P to contain a congruent copy of Q. Exactly the same problem may be asked for tetrahedra in three dimensions, though in this case the arrangement of the edges of different lengths is also relevant.

The most general problem would be to find an efficient algorithm that determines whether Q may be fitted inside P, given polygons (or polyhedra) P and Q described, say, by their vertex coordinates. Even more generally, find the similarity ratio by which Q must be contracted to fit inside P. To this end Hadwiger has given some sufficient conditions, in terms of length, area, etc., for one plane domain (not necessarily convex) to be able to contain another (see Santaló, Section I.7.6).

A. Bezdek & K. Bezdek, When is it possible to translate a convex polyhedron into another one? *Studia Sci. Math. Hungar.* **21** (1986) 337–342; *MR* **88j**:52014.

H. Hadwiger, Überdeckung ebener Bereiche durch Kreise und Quadrate, *Comment. Math. Helv.* **13** (1941) 195–200; *MR* **3**, 90.

H. Hadwiger, Gegenseitige Bedeckbarkeit zweier Eibereiche und Isoperimetrie, *Vierteljschr. Naturforsch. Ges. Zürich* **86** (1941) 152–156; *MR* **4**, 112.

L. A. Santaló, *Integral Geometry and Geometric Probability*, Addison-Wesley, Reading, Mass., 1976; *MR* **55** #6340.

H. Steinhaus, *One Hundred Problems in Elementary Mathematics*, Pergamon, Oxford, 1964, 98.

B2. **Inscribing polygons in curves.** Given a simple closed curve C in the plane (that is a continuous non-self-intersecting image of a circle) can we always find four points on C that are the vertices of a square, as in Figure B2? Some unconvincing "solutions" of this have appeared, but it has only been validly proved for curves which are sufficiently smooth (see Schnirelman or Jerrard and the recent paper by Stromquist, who shows that it is true for piecewise C^1 curves without cusps). In the special case where C is a convex curve, rather shorter proofs have been given by Emch, Zindler, and Christensen. Klee has suggested that the general case might be tackled using methods of nonstandard analysis. Can we always find such a square with the vertices occurring in the same order on the square and on C?

A very similar question, still open, was raised by Steinhaus and proved by him for the special case of convex curves: Given a simple closed curve C in

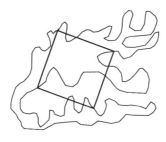

Figure B2.   Does every closed curve contain the vertices of some square?

the plane, do there always exist three concurrent chords bisecting each other and cutting at prescribed angles?

Suppose that C is a simple closed space curve (imagine a closed loop of wire). Greenspan asks whether C must contain either four concyclic points or three collinear points, and Hadwiger asks whether C always contains the vertices of a parallelogram. Is the problem easier if we assume that C is unknotted?

Does every closed space curve C contain the vertices of a quadrilateral with equal sides and equal diagonals? Stromquist has shown that this is so if C has a continuously turning tangent, but is it true in general?

At this point it is appropriate to mention Mizel's problem: is the circle the only closed convex curve such that whenever three vertices of a rectangle lie on it, so does the fourth? Several authors showed that this is the case, including Danzer, who asked if the result remained true for closed, nonconvex curves. This was recently established by Zamfirescu.

J. G. Ceder, A property of planar convex bodies, *Israel J. Math.* **1** (1963) 248–253; *MR* **29** #6391.

J. Ceder & B. Grünbaum, On inscribing and circumscribing hexagons. *Colloq. Math.* **17** (1967) 99–101; *MR* **35** #3544.

C. Christensen, A square inscribed in a convex figure, *Mat. Tidsskr.* **B** (1950) 22–26; *MR* **12**, 525.

L. W. Danzer, A characterization of the circle, in [Kle], 99–100; *MR* **27** #4141b.

A. Emch, Some properties of closed convex curves in a plane, *Amer. J. Math.* **35** (1913) 407–412.

A. Emch, On the medians of a closed convex polygon, *Amer. J. Math.* **37** (1915) 19–28.

D. Greenspan, Problem 4774, *Amer. Math. Monthly* **65** (1958) 125.

H. Guggenheimer, Finite sets on curves and surfaces, *Israel J. Math.* **3** (1965) 104–112; *MR* **32** #6326.

H. Hadwiger, Ungelöste Problem 53, *Elem. Math.* **26** (1971) 58; *MR* **48** #4398.

R. P. Jerrard, Inscribed squares in plane curves, *Trans. Amer. Math. Soc.* **98** (1961) 234–241; *MR* **22** #11354.

V. Klee, Some unsolved problems in plane geometry, *Math. Mag.* **52** (1979) 131–145; *MR* **80m**:52006.

L. G. Schnirelmann, On certain geometrical properties of closed curves, *Uspehi Matem. Nauk* **10** (1944) 34–44; *MR* **7**, 35.

H. Steinhaus, On chords of convex curves, *Bull. Acad. Polon. Sci. Cl. III* **5** (1957) 595–597; *MR* **19**, 573.

W. Stromquist, Inscribed squares and square-like quadrilaterals in closed curves, *Mathematika* **36** (1989) 187–197.

T. Zamfirescu, An infinitesimal version of the Besicovitch–Danzer characterization of the circle, *Geom. Dedicata* **27** (1988) 209–212; *MR* **89i**:52003.

K. Zindler, Über konvexe Gebilde, *Monatsh. Math.* **31** (1921) 25–57.

B3. Maximal regular polyhedra inscribed in regular polyhedra. Denote the regular polyhedra or Platonic solids by: tetrahedron T, cube C, octahedron O, dodecahedron D, and icosahedron I. What is the maximum of the ratio (volume of P)/(volume of Q) where P and Q are regular polyhedra of given types, with P contained in Q? Croft sets up a general theory for problems of this nature and obtains solutions for 14 of the 20 possible pairings. (Robbins also treats C in T.) The outstanding cases are the two reciprocal pairs C in I, D in O, and T in I, D in T, and the self-reciprocal cases D in I and I in D.

H. T. Croft, On maximal regular polyhedra inscribed in a regular polyhedron, *Proc. London Math. Soc.* (3) **41** (1980) 279–296; *MR* **81m**:51027.
D. P. Robbins, Solution to problem E2349, *Amer. Math. Monthly* **83** (1976) 54–58.

B4. **Prince Rupert's Problem.** How big a cube can be made to pass through a unit cube? This problem is attributed to Prince Rupert and the first solution attributed to Pieter Nieuwland. It may be reformulated as: How large a square may be inscribed in a unit cube? Then the cube having this as a base can tunnel through the unit cube in a direction perpendicular to the plane of the maximum square.

The four corners of this square are situated one quarter of the way along four edges of the cube as in Figure B3. The side of the square, and thus the edge of the cube, is $\frac{3}{4}\sqrt{2} = 1.0607\ldots$. The moving cube cuts the front edge $3/16$ of the way up, and the back edge $3/16$ of the way down.

Note that the four corners of the maximum square, together with the points $1/4$ of the way down the front edge and $1/4$ of the way up the back edge, form the vertices of the largest regular octahedron that can be inscribed in the cube (see Section B3).

More generally, we can ask how large an m-dimensional cube can be inscribed in a unit n-dimensional cube, where $m < n$. Let its edge be $f(m, n)$. Then $f(1, n) = \sqrt{n}$, and the above result may be written $f(2, 3) = \frac{3}{4}\sqrt{2}$. It is not hard to show that if $k \le m \le n - 1$, then

$$f(m + 1, n) < f(m, n) < f(m, n + 1) \text{ and } f(k, m)f(m, n) \le f(k, n).$$

Also $f(m, n) \le \sqrt{n/m}$, with equality just if m divides n. For squares in even-dimensional cubes $f(2, 2n) = \sqrt{n}$ but the best result for odd-dimensional cubes that we know is

$$\sqrt{(8n + 1)/8} \le f(2, 2n + 1) \le \sqrt{(2n + 1)/2},$$

and for larger m little seems to be known. Can the bound $f(3, 4) \le \frac{2}{3}\sqrt{3} = 1.1547\ldots$ be attained?

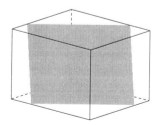

Figure B3.   The largest square that may be inscribed in a unit cube.

D. J. E. Schrek, Prince Rupert's problem and its extension by Pieter Nieuwland, *Scr. Math.* **16** (1950) 73–80.

D. Wells, *The Penguin Dictionary of Curious and Interesting Numbers*, Penguin Books, Harmondsworth, England, 1986, 33, where $3\sqrt{2}/4$ is misprinted as $3/4\sqrt{2}$.

B5. Random polygons and polyhedra. What is the expected volume of a tetrahedron with four vertices chosen at random (with uniform distribution) in a given tetrahedron K of unit volume? Since the problem is affinely invariant, we may assume that K is a regular tetrahedron. Originally Klee conjectured that the expected volume is $\frac{1}{60}$, but "experimental" results have suggested that the answer is nearer to $\frac{1}{57}$. It is easy to see that the problem is equivalent to finding the expectation of a certain determinant with random entries. However, all that seems to be known is that the expected volume of the tetrahedron is $\frac{2}{5}$ of the expected volume of the convex hull of a set of five random points.

This is the prototype for a host of problems in geometrical probability. If K is a given convex set of unit volume in $\mathbb{R}^d$, and n points of K are chosen at random, what are the mean, variance, and higher moments of the distribution of the area or volume, width, diameter, perimeter, or surface area, etc., of the convex hull of the n points? For which K are these quantities maximized or minimized?

There is a considerable literature on these problems, particularly for the plane case. Let $V(K, n)$ denote the expected area (or volume in higher dimensions) of the convex hull of n random points in a convex set K of unit area (or volume). Many years ago, Blaschke, Crofton, and Czuber established that for plane K

$$35/48\pi^2 = V(\text{ellipse}, 3) \le V(K, 3) \le V(\text{triangle}, 3) = \tfrac{1}{12}.$$

Alikaski proved that

$$V(\text{regular } m\text{-gon}, 3) = (9\cos^2\omega + 52\cos\omega + 44)/36m^2\sin^2\omega,$$

where $\omega = 2\pi/m$. For larger n, Rényi & Sulanke showed that if the boundary of K has continuously varying curvature then

$$V(K, n) = 1 - cn^{-2/3} + O(1/n),$$

where c may be computed from the geometry of K. They obtained a similar result for expected perimeter length. On the other hand, Buchta showed that

$$V(\text{triangle}, n) = 1 - \frac{2}{n+1}\sum_{k=1}^{n}\frac{1}{k},$$

and obtained a similar expression for $V(\text{parallelogram}, n)$. Affentranger has obtained a (rather unwieldy) recurrence relation for $V(K, n)$ for arbitrary K.

A number of authors have considered the distribution of the "shapes" of random traingles [see Kendall (& Le) for further details and references].

The higher dimensional analogs of such problems are much harder. Groemer proved that in $\mathbb{R}^d$, the expected (d-dimensional) volume $V(K, d + 1)$ of a random simplex in a convex set K of unit volume, is maximal when K is an ellipsoid, and Kingman calculated the maximum values. Buchta established that

$$V(\text{tetrahedron}, n) \sim 1 - \tfrac{3}{4}(\log n)^2/n$$

as $n \to \infty$, and Dwyer showed that

$$1 - V(K, n) \leq c(\log n)^{d-1}/n$$

for any polytope K in $\mathbb{R}^d$. Recently Bárány & Larman proved a remarkable result relating the expected volume directly to the geometry of K. For $\varepsilon > 0$ write $K(\varepsilon)$ for the set of $\mathbf{x}$ in K such that there is a (hyper-) plane through $\mathbf{x}$ cutting a volume of ε or less off K. [Thus, $K(\varepsilon)$ is a "boundary shell" of K.] Then there are positive constants c_1, and c_2 such that

$$c_1 \, \text{vol} \, K(1/n) \leq 1 - V(K, n) \leq c_2 \, \text{vol} \, K(1/n).$$

Bárány has gone on to develop results of this nature for other measurements of the convex hull of n random points in K.

A closely related question, often known as Sylvester's problem, asks for the probability that four points chosen at random in a convex region K form a convex quadrilateral (i.e., with none of the points inside the triangle formed by the other three). It is easy to see that the probability is $1 - 4V(K, 3)/A$, where K has area A and $V(K, 3)$ is the mean area of random triangles in K defined above (see Kendall & Moran). The generalizations of this provide some challenging problems: if n points are chosen at random in a region K in $\mathbb{R}^d$, what is the probability that exactly k of the points lie on the boundary of the convex hull of the n points, for $0 \leq k \leq n - d - 1$? (See Groeneboom.)

For another variation, suppose two random triangles are found in a plane convex set K. What is the probability that they are disjoint, and what is the probability that one contains the other? If two such triangles are chosen in a 3-dimensional set, what is the probability that they are linked?

If five points are chosen at random in K, what is the probability that the (almost surely) unique conic section through the points is an ellipse rather than a hyperbola?

Chernoff asks about qualitative properties of random polygons when the vertices are ordered. Let $\mathbf{x}_1$, $\mathbf{x}_2$, ..., $\mathbf{x}_n$ be n random points selected independently in the unit disk, say. What is the probability that the polygon $\mathbf{x}_1$, $\mathbf{x}_2$, ..., $\mathbf{x}_n$, $\mathbf{x}_1$ is (a) non-self-intersecting, and (b) convex? If the $\mathbf{x}_i$ are randomly selected from a subset of $\mathbb{R}^3$, what is the probability that the polygonal curve obtained is knotted, and what is the probability of each particular type of knot?

Buchta & Tichy have recently investigated analogous problems on general surfaces and manifolds, and have obtained some results in the case of the d-dimensional torus.

There is no end to problems of this nature. The tract by Kendall & Moran
is a good introduction to this fascinating area of geometric probability.

F. Affentranger, Generalization of a formula of C. Buchta about the convex hull of
    random points, *Elem. Math.* **43** (1988) 39–45, 151–152; *MR* **89g**:52003.
H. A. Alikaski, Über das Sylvestische Vierpunktproblem, *Ann. Acad. Sci. Fenn A* **51**
    (1939) 1–10.
A. Baddeley, A fourth note on recent research in geometric probability, *Adv. Appl.
    Probab.* **9** (1977) 824–860; *MR* **57** #7724.
I. Bárány, Intrinsic volumes and *f*-vectors of random polytopes, *Math. Ann.* **285** (1989)
    671–699.
I. Bárány & Z. Füredi, On the shape of the convex hull of random points, *Probab.
    Theory Related Fields* **77** (1988) 231–240; *MR* **89g**:60030.
I. Bárány & D. G. Larman, Convex bodies, economic cap covering, random polytopes,
    Mathematika **35** (1988) 274–291.
W. Blaschke, Über affine Geometrie XI: Lösung des "Vierpunktproblems" von Sylves-
    ter aus der Theorie der geometrischen wahrscheinlichkeiten, *Leipziger Berichte*
    69 (1917) 436–453.
C. Buchta, Zufallspolygone in konvexen Vielecken, *J. Reine Angew. Math.* **347** (1984)
    212–220; *MR* **85e**:60009.
C. Buchta, Das Volumen von zufallspolyedern im Ellipsoid, *Anz. Österreich Akad.
    Wiss. Math.-Natur. Kl.* **121** (1984) 1–5; *MR* **87e**:60025.
C. Buchta, A note on the volume of a random polytope in a tetrahedron, *Illinois J.
    Math.* **30** (1986) 653–659; *MR* **87m**:60038.
C. Buchta, On a conjecture of R. E. Miles about the convex hull of random points,
    Monatsh. Math. **102** (1986) 91–102; *MR* **88f**:60016.
C. Buchta, Distribution independent properties of the convex hull of random points,
    J. Theor. Probab. **3** (1990) 387–393.
C. Buchta & R. F. Tichy, Random polytopes on the torus, *Proc. Amer. Math. Soc.* **93**
    (1985) 312–316; *MR* **86c**:60021.
P. R. Chernoff, Some problems in geometric probability, *Amer. Math. Monthly* **90**
    (1983) 120.
M. W. Crofton, Probability, *Encyclopedia Brittanica*, 9th ed. **19** (1885) 768–788.
E. Czuber, *Wahrscheinlich und ihre Anwedung auf Fehlerausgleichung Statistik, und
    Lebensversicherung,* 1, Teubrer, Leipzig, 1903.
R. A. Dwyer, On the convex hull of random points in a polytope, *J. Appl. Probab.* **25**
    (1988) 688–699.
Z. Füredi, Random polytopes in the *d*-dimensional cube, *Discrete Comput. Geom.* **1**
    (1986) 315–319; *MR* **88d**:52003.
H. Groemer, On some mean values associated with a randomly selected simplex in a
    convex set, *Pacific J. Math.* **45** (1973) 525–533; *MR* **47** #5916.
H. Groemer, On the mean value of a random polytope in a convex set, *Arch. Math.*
    (*Basel*) **25** (1974) 86–90; *MR* **49** #6036.
H. Groemer, On the average size of polytopes in a convex set, *Geom. Dedicata* **13** (1982)
    47–62; *MR* **84d**:52013.
P. Groeneboom, Limit theorems for convex hulls, *Probab. Theory Related Fields* **79**
    (1988) 327–368; *MR* **89j**:60024.
P. M. Gruber, Approximation of convex bodies, in [GW], 131–162; *MR* **85d**:52001.
G. R. Hall, Acute triangles in the *n*-ball, *J. Appl. Probab.* **19** (1982) 712–715.
N. Henze, Random triangles in convex regions, *J. Appl. Probab.* **20** (1983) 111–125;
    MR **84g**:60019.
M. G. Kendall, Exact distributions for the shape of random triangles in convex sets,
    Adv. Appl. Probab. **17** (1985) 308–329; *MR* **86m**:60033.

M. G. Kendall & H.-L. Le, Exact shape densities for random triangles in convex polygons, *Adv. Appl. Probab.* **1986**, suppl., 59–72; *MR* **88h**:60022.

M. G. Kendall & P. A. P. Moran, *Geometric Probability*, Hafner, New York, 1963; *MR* **30** #4275.

J. F. C. Kingman, Random secants of a convex body, *J. Appl. Probab.* **6** (1969) 660–672; *MR* **40** #8098.

V. Klee, What is the expected volume of a simplex whose vertices are chosen at random from a given convex body? *Amer. Math. Monthly* **76** (1969) 286–288.

H.-L. Le, Explicit formulae for polygonally generated shape-densities in the basic tile, *Math. Proc. Cambridge Philos. Soc.* **101** (1987) 313–321; *MR* **88f**:60019.

H.-L. Le, Singularities of convex-polygonally generated shape-densities, *Math. Proc. Cambridge Philos. Soc.* **102** (1987) 587–596; *MR* **88k**:60024.

W. Moser & J. Pach, Problem 35, [MP].

R. E. Pfiefer, The historical development of J. J. Sylvester's four point problem, *Math. Mag.* **62** (1989) 309–317; *MR* **90m**:52001.

W. J. Reed, Random points in a simplex, *Pacific J. Math.* **54** (1974) 183–198; *MR* **51** #8959.

A. Rényi & R. Sulanke, Über die konvexe Hülle von n zufällig gewählten Punkten, I. II, *Z. Wahrschein. Verw. Gebiete* **2** (1963) 75–84, **3** (1964) 138–147; *MR* **27** #6190; **29** #6392.

L. A. Santaló, *Integral Geometry and Geometric Probability*, Addison-Wesley, Reading, Mass., 1976; *MR* **55** #6340.

R. Schneider, Approximation of convex bodies by random polytopes, *Aequationes Math.* **32** (1987) 304–310; *MR* **88j**:52012.

R. Schneider & J. A. Wieacker, Random polytopes in a convex body, *Z. Wahrschein. Verw. Gebiete* **52** (1980) 69–73; *MR* **81f**:52008.

D. Stoyan, W. S. Kendall & J. Mecke, *Stochastic Geometry and its Applications*, Wiley, Chichester, 1987; *MR* **88j**:60034.

J. J. Sylvester, On a special class of questions on the theory of probabilities, *British Assoc. Report* (1865) 8–9.

B6. **Extremal problems for polygons.** What is the greatest possible area of an n-sided convex polygon of unit diameter? When asking this question Reinhardt showed that for odd n the extremum is given by the regular polygon. For even n it is *not*, in general, the regular polygon. If $n = 4$ there is an infinite class of quadrilaterals, all with perpendicular diagonals of unit length, that give maximum area $\frac{1}{2}$, equal to that of the square. Graham has found the extremum for $n = 6$ which has area about 4% greater than the regular hexagon of unit diameter. For even $n > 6$ the problem is open, as are the higher dimensional analogs, except in very special cases.

Similarly, what is the maximum perimeter length of a convex n-gon of diameter 1? Provided that n has an odd factor, the extremal is given by certain n-gons inscribed in regular Reuleaux p-gons of (constant) width one, where p is an odd factor of n, the vertices of the Reuleaux p-gon being among those of the n-gon. Even if n is a power of 2, the n-gon of maximum perimeter is not regular.

More generally, one can restrict any extremal problem for convex sets to the class of convex n-gons. If this can be solved for each n, an approximation method often leads to a solution to the problem for the class of all convex sets. With this idea in mind, Rogalski & Saint-Raymond ask about the minimum of $A(\frac{1}{2}L - D)^{-2}$ over centro-symmetric n-gons.

H. Bieri, Zweiter Nachtrag zu nr. 12, *Elem. Math.* **16** (1961) 105–106.

R. L. Graham, The largest small hexagon, *J. Combin. Theory Ser. A* **18** (1975) 165–170; *MR* **50** #12803.

H. Hadwiger, Ungelöste Probleme 12, *Elem. Math.* **11** (1956) 86.

B. Kind & P. Kleinschmidt, On the maximal volume of convex bodies with few vertices, *J. Combin. Theory Ser. A* **21** (1976) 124–128; *MR* **53** #11500.

D. G. Larman & N. K. Tamvakis, The decomposition of the n sphere and the boundaries of plane convex domains, in [RZ], 209–214; *MR* **87b**:52032.

H. Lenz, Zerlegung ebener Bereiche in konvexe Zellen von möglichst kleinem Durchmesser, *Jber. Deutsch. Math. Verein.* **58** (1956) 87–97; *MR* **18**, 817.

K. Reinhardt, Extremale polygone mit gegebenen Durchmessers, *Jber. Deutsch. Math. Verein.* **31** (1972) 251–270.

M. Rogalski & J. Saint-Raymond, Inequalities about symmetric compact convex sets in the plane, *Amer. Math. Monthly* **92** (1985) 466–480; *MR* **86k**:52013.

J. J. Schäffer, Nachtrag zu nr. 12, *Elem. Math.* **13** (1957) 85–86.

N. K. Tamvakis, On the perimeter and the area of the convex polygons of a given diameter, *Bull. Soc. Math. Grèce (N.S.)* **28** (1987) 115–132; *MR* **89g**:52008.

S. Vincze, On a geometrical extremum problem, *Acta Sci. Math. (Szeged) A* **12** (1950) 136–142; *MR* **12**, 352.

B7. **Longest chords of polygons.** Let P be a plane convex n-gon with side lengths $a_1, \ldots, a_n$. Let b_i be the length of the longest chord of P parallel to the ith side. Fejes Tóth showed that $\sqrt{8} < \sum_{i=1}^{n} a_i/b_i \leq 4$, with equality on the right if and only if P is a parallelogram. He conjectures that $3 \leq \sum_{i=1}^{n} a_i/b_i$ with equality only for a snub triangle obtained by cutting off three congruent triangles from the corners of a triangle.

What are the 3-dimensional analogs of these inequalities for areas of plane sections parallel to the faces of polyhedra?

L. Fejes Tóth, Über eine affineinvariante Masszahl bei Eipolyedern, *Studia Sci. Math. Hungar.* **5** (1970) 173–180; *MR* **42** #3673.

B8. **Isoperimetric inequalities for polyhedra.** What is the minimum of the **isoperimetric quotient** S^3/V^2 over all polyhedra with f faces? Goldberg conjectures that the minimum is always attained by a **medial** polyhedron, that is one in which all the faces are either $\lfloor 6 - 12/f \rfloor$-gons or $(\lfloor 6 - 12/f \rfloor + 1)$-gons. (Medial polyhedra do not exist if $f = 11$ or 13, so the conjecture does not apply in this case. For other values of f with $4 \leq f \leq 15$ there is exactly one combinatorial class of medial polyhedra.) In particular the regular tetrahedron, the cube, and the regular dodecahedron are minimal for $f = 4, 6$, and 12; in fact for any polyhedron with f faces

$$S^3/V^2 \geq 54(f - 2)(4 \sin^2 \alpha_f - 1) \tan \alpha_f,$$

where $\alpha_f = \pi f/6(f - 2)$, with equality just in these three cases (see Section V7 of Fejes Tóth).

The regular octahedron does **not** give the minimum of S^3/V^2 among the 8-faced polyhedra. However it is the minimal polyhedron within its combinatorial class. An old conjecture of Steiner is that the Platonic solids minimize S^3/V^2 among all polyhedra of the same class. This remains unproved in the case of the icosahedron.

Which combinatorial classes are "stable" in the sense that the minimum is attained strictly within the class?

L. Fejes Tóth, The isepiphan problem for n-hedra, *Amer. J. Math.* **70** (1948) 174–180; *MR* **9**, 460.
L. Fejes Tóth, [Fej].
M. Goldberg, The isoperimetric problem for polyhedra, *Tôhoku Math. J.* **40** (1935) 226–236.
J. Steiner, Sur le maximum et le minimum des figures, *J. Math. Berlin* **24** (1842) 93–152, 189–250.
G. Weiss, On isoperimetric tetrahedra, in [BF], 631–637; *MR* **88m**:52022.

B9. **Inequalities for sums of edge lengths of polyhedra.** Let P be a convex polyhedron in $\mathbb{R}^3$ with volume V, surface area A, and sum of edge lengths L. Aberth proves that

$$A/L^2 < (6\pi)^{-1}.$$

The constant $(6\pi)^{-1}$ can certainly be reduced—what is the least possible value? Similarly, Melzak conjectures that

$$V/L^3 \le (2^4 3^{11})^{-1/6}$$

with equality only for a right prism based on an equilateral triangle of side length equal to the height of the prism. The problem becomes much easier if restricted to tetrahedra. Here the extremals have been shown to be regular in the volume case by Melzak and in the surface area case by Kömhoff.

More generally, if P is a d-polytope, with $d \ge 3$, let V_k be the k-dimensional content of the k-dimensional faces of P. Thus, V_0 is the number of vertices, V_1 the sum of edge lengths, V_2 the sum of the areas of the 2-dimensional faces, ..., V_{d-1} the $(d-1)$-dimensional surface area, and V_d the volume of P. For each $1 \le i, j \le d$ let $c(d, i, j)$ be the smallest number such that for all d-polytopes P

$$V_i^{1/i}/V_j^{1/j} \le c(d, i, j).$$

Find upper estimates for, and optimistically the actual value of, $c(d, i, j)$. As Klee points out, it is far from clear whether these ratios are even bounded, and the first step should be to establish which $c(d, i, j)$ are finite! Eggleston, Grünbaum & Klee show that $c(d, i, j)$ is finite if $i = d$ or $i = d - 1 > j$, or i is a multiple of j, and Klee points out that $c(d, i, j) = \infty$ if $i < j$; it seems likely that $c(d, i, j)$ is finite for all other cases, but this is not known if $d - 2 \ge i > j \ge 2$ and i is not a multiple of j. The supremum $c(d, d, d - 1)$ is finite but not attained by any polytope, but Grünbaum conjectures that all other finite bounds are attained.

O. Aberth, An isoperimetric inequality for polyhedra and its application to an extremal problem, *Proc. London Math. Soc.* (3) **13** (1963) 322–336; *MR* **26** #6861.

H. G. Eggleston, B. Grünbaum & V. Klee, Some semicontinuity theorems for convex polytopes and cell complexes, *Comment. Math. Helv.* **39** (1964) 165–188; *MR* **30** #5217.

V. Klee, Which isoperimetric ratios are bounded? *Amer. Math. Monthly* **77** (1970) 288–289.

M. Kömhoff, An isoperimetric inequality for convex polyhedra with triangular faces, *Canad. Math. Bull.* **11** (1968) 723–727; *MR* **39** #2067.

M. Kömhoff, On a 3-dimensional isoperimetric problem, *Canad. Math. Bull.* **13** (1970) 447–449; *MR* **43** #2610.

Z. A. Melzak, Problems connected with convexity, Problem 13, *Canad. Math. Bull.* **8** (1965) 565–573; *MR* **33** #4781.

Z. A. Melzak, An isoperimetric inequality for tetrahedra, *Canad. Math. Bull.* **9** (1966) 667–669; *MR* **35** #879.

B10. Shadows of polyhedra.

Let P be a 3-dimensional convex polyhedron with n vertices. Consider the "shadow" (orthogonal projection) of P on the plane H. This will be a convex polygon with $f(P, H)$ vertices, say. For each n estimate

$$f(n) = \min_{P} \max_{H} f(P, H), \qquad g(n) = \max_{P} \min_{H} f(P, H);$$

in particular, how do they behave asymptotically as $n \to \infty$?

Moser, to whom this problem is due, conjectures that $f(n) = O(\ln n)$. If true, this is the largest possible estimate: take a regular tetrahedron, add vertices a distance ε outwards from the centroid of each face, then vertices ε^2 away from the centroid of each new face, and so on.

We conjecture that $g(n) = kn + O(1)$, perhaps even with $k = 1$. A case of interest is when P is a **cyclic** polyhedron, that is, the convex hull of a set of points all on the "moment curve" $x = t$, $y = t^2$, $z = t^3$.

A polyhedron P is called **equiprojective** if, for some k, the shadow of P in every direction (except those directions parallel to faces of P) is a k-gon. For example, a cube is equiprojective with almost all projections hexagonal. Shephard asks for a description of all equiprojective polyhedra. For which k do they exist?

A sort of dual to these problems concerns sections of polyhedra. Let $h(P, H)$ be the number of sides of the polygon formed by the intersection of the n-vertex polyhedron P with the plane H. For each n, what is

$$h(n) = \min_{P} \max_{H} h(P, H)?$$

Shephard asks if $h(n) \geq \sqrt{n}$ for $n \geq 4$.

Related results on the number of vertices of projections and sections of polyhedra and polytopes may be found in the papers by Bol and Klee.

G. Bol. Über Eikörper mit Vieleckschatten, *Math. Z.* **48** (1942) 227–246; *MR* **5**, 10.

V. Klee, Some characterizations of convex polyhedra, *Acta Math.* **102** (1959) 79–107; *MR* **21** #4390.

V. Klee, On a conjecture of Lindenstrauss, *Israel J. Math.* **1** (1963) 1–4; *MR* **28** #523.

G. C. Shephard, Twenty problems on convex polyhedra, II, *Math. Gaz.* **52** (1968) 359–367.

B11. **Dihedral angles of polyhedra.** Let P be a convex polyhedron. Suppose every dihedral angle of P is at most α, where $0 < \alpha < \pi$. (The **dihedral angle** between two adjacent faces is the internal angle between the two faces.) Estimate the maximum number of faces, edges, and vertices that P can have. In particular at what rate do these numbers tend to infinity as α approaches π?

Z. A. Melzak, More problems connected with convexity, Problem 3, *Canad. Math. Bull.* **11** (1968) 482–494; *MR* **38** #3767.

B12. **Monostatic polyhedra.** A **monostatic** or **unistable** polyhedron is one which, when made of material of uniform density, will rest in equilibrium on only one of its faces. Conway has shown that no tetrahedron can be monostatic, but Guy and Knowlton independently discovered a monostatic 19-faced solid; moreover, Guy describes such an object with 21 faces with the "bottom" as the face of least diameter and area. What is the least number of faces a monostatic polyhedron can have? Heppes discovered a "two-tip tetrahedron" which Guy realized with edges 41, 26, 24, 20, 17, and 4, with opposite pairs summing to 45, 43, and 44. (Amateur carpenters may like to construct this!)

Conway also asks the following:

(a) Can a monostatic polyhedron have n-fold symmetry for any $n > 2$?
(b) Which convex bodies can be approximated arbitrarily closely by monostatic polyhedra, in particular what about the sphere?
(c) How small can the diameter/girth ratio of such a solid be? Certainly $3/\pi + \varepsilon$ is possible. (The girth is the minimum length of the perimeter of a projection onto a plane.)

In higher dimensions, Dawson showed that no monostatic simplex exists in 6 or fewer dimensions, and also exhibited a 10-dimensional monostatic simplex. What is the smallest dimension in which such a simplex can exist?

J. H. Conway & R. K. Guy, Stability of polyhedra, *SIAM Rev.* **11** (1969) 78–82.
R. J. MacG. Dawson, Monostatic simplexes, *Amer. Math. Monthly* **92** (1985) 541–546; *MR* **86m**:52007.
R. K. Guy, Twenty odd questions in combinatorics, in *Combinatorial Mathematics and its Applications*, R. C. Bose et al. (eds.), Chapel Hill N. C., 1970, 209–237; *MR* **42** #1666.
A. Heppes, A double tipping tetrahedron, *SIAM Rev.* **9** (1967) 599–600.

B13. **Rigidity of polyhedra.** Early in the 19th century, Cauchy proved that any convex polyhedron in $\mathbb{R}^3$ with rigid faces, but hinged at the edges, was in fact completely rigid (i.e., it could not be subjected to small perturbations). Steinitz refined Cauchy's proof to show that a convex polyhedron was completely determined given its faces and a knowledge of which of their edges were

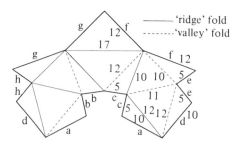

Figure B4.   A net for Steffen's construction of a non-rigid polyhedron of 9 vertices and 14 faces. The edge lengths have vertical symmetry, and the outer edges are joined as indicated by the letters.

joined. The "rigidity conjecture" that any (not necessarily convex) polyhedron homeomorphic to a sphere was rigid was disproved by Connelly, who described a simple example of an 18-faced, 11-vertex polyhedron that may be "flexed."

There are still many open problems; Connelly has written several good surveys. What is the minimum number of faces and vertices that a counterexample to the rigidity conjecture can have? The best example to date, due to Steffen, may be constructed from the "net" of Figure B4 and has 14 faces and 9 vertices. Connelly's Mathematical Intelligencer and Mathematics Magazine articles give constructional details of other flexible polyhedra.

Does every non-rigid polyhedron have constant volume under flexing, as do all known examples? This is the "bellows conjecture," that one can not construct a polyhedral bellows. Alexander shows that the integral mean curvature is constant. It is further conjectured that any two positions of a flexible polyhedron are equidecomposable into a finite number of polyhedra, and thus satisfy the Dehn–Sydler conditions (see Section C20). There are related questions for surfaces. Are there any flexible smooth (class C^2) surfaces in $\mathbb{R}^3$ (perturbations must preserve the surface metric)?

R. Alexander, Lipschitzian mappings and total mean curvature of polyhedral surfaces I, *Trans. Amer. Math. Soc.* **288** (1985) 661–678; *MR* 86c:52004.

A. L. Cauchy, Sur les polygones et les polyèdres, Second Mémoire I. École Polytech. **9** (1913) 87.

R. Connelly, A flexible sphere, *Math. Intelligencer* **1** (1978) 130–131; *MR* **58** #13055.

R. Connelly, A counter example to the rigidity conjecture for polyhedra, *Publ. Math. I.H.E.S.* **47** (1977) 333–338; *MR* **58** #7642.

R. Connelly, The rigidity of polyhedral surfaces, *Math. Mag.* **52** (1979) 275–283; *MR* **80k**:53089.

R. Connelly, Conjectures and open questions in rigidity, Proc. Int. Congress Math. (Helsinki 1978), Acad. Sci. Fennica, Helsinki, 1980, 407–414; *MR* **82a**:51009.

R. Connelly, Flexing surfaces, [Kla], 79–89.

R. Connelly, Rigidity and energy, *Invent. Math.* **66** (1982) 11–33; *MR* **83m**:52012.

H. Gluck, Almost all simply connected closed surfaces are rigid, in *Geometric Topology, Lecture Notes in Mathematics*, **438**, 225–239, Springer-Verlag, Berlin, 1975; *MR* **53** #4074.

E. Steinitz, Polyeder und Raumeneinteilungen, *Encykl. Math. Wiss.* 3 (1922) Geometrie, Pt.3 **12** (1916) 1–139.

B14. **Rigidity of frameworks.** A **framework** in $\mathbb{R}^d$ consists of a configuration of vertices some pairs of which are joined by rigid "rods." A framework is **(infinitesimally) rigid** if there are no small perturbations that preserve the lengths of all the rods (other than congruences of the whole structure). Thus, in the plane or in space, a triangle is rigid, but a quadrilaterial is not.

We seek necessary and sufficient conditions for a framework to be **minimally rigid**, that is for the framework to be rigid, but such that it ceases to be so if any rod is removed. Laman and Assinov & Roth have shown that a framework in the plane with v vertices is minimally rigid if and only if there are $2v - 3$ rods arranged in such a way that, for $2 \leq v' \leq v$, every subset of v' of the vertices are joined by at most $2v' - 3$ of the rods.

The 3-dimensional situation is much more complicated. A minimally rigid framework with v vertices must have $3v - 6$ rods, with every subset of v' vertices joined by at most $3v - 6$ rods for $2 \leq v' \leq v$, but this condition is not sufficient for minimal rigidity. Precise conditions are known for the rigidity of certain classes of framework (e.g., for complete bipartite frameworks where the vertices are divided into two sets with rods joining each vertex in one set to each vertex in the other, see Whiteley). Can necessary and sufficient conditons be found for the rigidity of general frameworks in $\mathbb{R}^3$? Kahn has shown that there is an algorithm that will determine in a finite time whether a given framework is rigid.

Another situation that has been completely analyzed is where the vertices and rods of a framework are the vertices and edges of a convex polyhedron. Suppose that such a framework has e rods and v vertices, the polyhedron has $2 + e - v$ faces by Euler's relationship. It follows from Cauchy's rigidity theorem (see Section B13) that if the faces are all triangular, that is, if $3v = 6 + e$, then the structure is rigid and not susceptible to small perturbations. Roth has proved that this condition is also sufficient: if one of the faces is not triangular, then the framework cannot be rigid.

Suppose that we wish to make a given convex polyhedral framework P rigid—such problems are frequently encountered in structural engineering. There are a number of ways of doing this.

(a) by the addition of rods between any vertices of P
(b) by the addition of rods between any vertices of P belonging to the same faces
(c) by the addition of rods joining any points on edges of the framework P
(d) by the addition of rods joining any points on edges of the same face of P
(e) to (h) the same as in (a) to (d) but using strings rather than rods
(i) to (l) the same as in (a) to (d) but using "struts" which support compression but not tension.

In each case we want to know what is the greatest number of rods, strings, or struts that are ever needed to make rigid any convex polyhedral framework of e edges and v vertices. Also which frameworks of e edges and v vertices require fewest additions to ensure rigidity?

Despite the amount that has been written on structural rigidity, numerical estimates are scarce. However, Whiteley has shown, by using $k - 2$ strings to join vertices of each k-gonal face when $k > 3$, that $2v - 4 - f_3$ strings are enough to ensure rigidity (f_3 is the number of triangular faces). Moreover, in the case of strings joining vertices of faces, this number is required for any convex polyhedral framework.

One can ask similar questions about making non-polyhedral frameworks rigid, though it seems hard to be systematic about such problems.

This subject is a vast one. Many related papers may be found in the engineering literature and in the journal "Structural Topology."

L. Asimov & B. Roth, Rigidity of graphs, *Trans. Amer. Math. Soc.* **245** (1978) 279–289; *MR* **80i**:57004a.

L. Asimov & B. Roth, Rigidity of graphs II, *J. Math. Anal. Appl.* **68** (1979) 171–190; *MR* **80i**:57004b.

E. D. Bolker, Bracing rectangular frameworks I, *SIAM J. Appl. Math.* **36** (1979) 491–508; *MR* **81j**:73066b.

E. D. Bolker & H. Crapo, Bracing rectangular frameworks I, *SIAM J. Appl. Math.* **36** (1979) 473–490; *MR* **81j**:73066a.

E. D. Bolker & B. Roth, When is a bipartite graph a rigid framework? *Pacific J. Math.* **90** (1980) 27–44; *MR* **82c**:57003.

R. Connelly, The rigidity of certain cabled frameworks and the second-order rigidity of arbitrarily triangulated convex surfaces, *Adv. Math.* **37** (1980) 272–299; *MR* **85a**:53059.

R. Connelly, Rigidity and energy, *Invent. Math.* **66** (1982) 11–33; *MR* **83m**:52012.

R. Connelly & M. Terrell, Globally rigid symmetric tensegrities, to appear.

H. Crapo & W. Whiteley, Statics of frameworks and motions of panel structures, a projective geometric introduction, *Struct. Topology* (1982) no. 6, 43–82; *MR* **84b**:51029.

H. Crapo & W. Whiteley (eds.), *The Geometry of Rigid Structures*, to appear.

P. Kahn, Counting types of rigid frameworks, *Invent. Math.* **55** (1979) 297–308; *MR* **80m**:51008.

G. Laman, On graphs and rigidity of plane structures, *J. Eng. Math.* **4** (1970) 331–340.

B. Roth, Rigid and flexible frameworks, *Amer. Math. Monthly* **88** (1981) 6–21; *MR* **83a**:57027.

B. Roth & W. Whiteley, Tensegrity frameworks, *Trans. Amer. Math. Soc.* **265** (1981) 419–446; *MR* **82m**:51018.

N. L. White & W. Whiteley, The algebraic geometry of stresses in frameworks, *SIAM J. Algebraic Discrete Methods* **4** (1983) 481–511; *MR* **85f**:52024.

N. L. White & W. Whiteley, The algebraic geometry of motions of bar-and-body of frameworks, *SIAM J. Algebraic Discrete Methods* **8** (1987) 1–32; *MR* **88m**:52016.

W. Whiteley, Motions and stresses of projected polyhedra, *Struct. Topology* (1982) no. 7, 13–38; *MR* **85h**:52010.

W. Whiteley, Rigidity and polarity. I. Statics of sheet structures, *Geom. Dedicata* **22** (1987) 329–362; *MR* **88h**:51020.

W. Whiteley, Infinitesimally rigid polyhedra, I, II, *Trans. Amer. Math. Soc.* **285** (1984) 431–465; **306** (1988) 115–139; *MR* **86c**:52010, **90a**:52014.

W. Whiteley, Infinitesimal motions of a bipartite framework, *Pacific J. Math.* **11** (1984) 233–255; *MR* **86c**:52009.

W. Whiteley, The projective geometry of rigid frameworks, in *Finite Geometries*, Lecture Notes in Pure Appl. Math. 103, 353–370, Marcel Dekker, New York, 1985; *MR* **87k**:52029.

B15.   Counting polyhedra. How many different convex polyhedra are there with v vertices and f faces (and thus by Euler's formula $v + f - 2$ edges)? By "different" we of course mean combinatorially different. The two polyhedra shown in Figure B5 each have 6 vertices, 12 edges, and 8 triangular faces, but they are not combinatorially equivalent. In the first polyhedron the faces meet four at a vertex; in the second there are vertices where as many as five, or as few as three, faces meet. By Steinitz's theorem the problem reduces to one of counting planar 3-connected graphs. Various computer algorithms have been developed to enumerate polyhedra with given numbers of vertices and faces, but the computer time involved is considerable and exact values have only been fixed for relatively small v and f. Enumeration of smaller polyhedra goes back to Euler, Steiner, Kirkman, and Hermes. Table B1 combines the results of Duijvestijn & Federico with those of Engel. Notice the symmetry between v and f; this is due to the fact that for a polyhedron P with v vertices and f faces and any origin interior to P, the polar reciprocal body $P' = \{\mathbf{x} : \mathbf{x} \cdot \mathbf{y} \leq 1 \text{ for all } \mathbf{y} \text{ in } P\}$ has f vertices and v faces. For all such origins the polyhedra P' are in the same combinatorial class, which may be regarded as dual to that of P.

It would, of course, be of greater interest to find a significantly more efficient enumeration algorithm than just to use variants of those already available to extend the table. It would also be of interest to find the computational complexity of the problem (i.e., to estimate the smallest number of computational steps demanded by the intrinsic nature of the problem for large v and f). How do the known algorithms compare with this? Linial has recently shown that such problems are $\#P$ complete.

If exact numbers cannot be found, we look for asymptotic estimates for the number of convex polyhedra. It has turned out to be rather easier to count "rooted" polyhedra, that is, with a particular vertex identified as a base point. However, it has recently been established that "most" polyhedra are asymmetric (in the combinatorial sense), so that the number of distinct convex

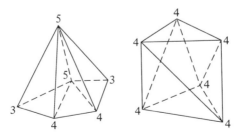

Figure B5.   Two polyhedra, both with 6 vertices, 12 edges, and 8 faces, that are not combinatorially equivalent.

Unsolved Problems in Geometry

Table B1. The numbers of combinatorially different polyhedra with given numbers of vertices and faces.

Faces→ Vertices↓	4	5	6	7	8	9	10	11	12
4	1								
5		1	1						
6		1	2	2	2				
7			2	8	11	8	5		
8			2	11	42	74	76	38	14
9				8	74	296	633	768	558
10				5	76	633	2635	6134	8822
11					38	768	6134	25626	64439
12					14	558	8822	64439	268394
13						219	7916	104213	706770*
14						50	4442	112082	1259093*
15							1404	79773	1552824*
16							233	36528	1334330*
17								9714	786625*
18								1249	304087*
19									69564*
20									7595
Total:	1	2	7	34	257	2606	32300	440564	6363115*

*approximate values only.

polyhedra can be established by dividing the number of rooted polyhedra with e edges by $4e$.

Thus the work of Bender & Richmond on rooted polyhedra, together with results of Bender & Wormald on symmetry have recently confirmed the asymptotic estimate for the number of polyhedra with v vertices:

$$\sim \left(\frac{4+\sqrt{7}}{64\pi\sqrt{7}}\right)^{1/2}\left(\frac{3+\sqrt{7}}{7+\sqrt{7}}\right)\left(\frac{38\sqrt{7}-100}{9}\right)^{5/2} v^{-7/2}\left(\frac{16(17+7\sqrt{7})h}{27}\right)^{v-1}.$$

Of course, using duality, one gets the asymptotic number of polyhedra with f faces by replacing v by f in the above formula. The asymptotic formula for simplicial polytopes obtained by Tutte is simpler:

$$\sim \frac{1}{8}\left(\frac{1}{3\pi}\right)^{1/2} v^{-7/2}\left(\frac{256}{27}\right)^{v/2} \qquad (v \text{ even}).$$

Tutte "nearly" proved that the number of convex polyhedra with e edges is asymptotic to

$$\frac{e^{-7/2}4^e}{486\sqrt{\pi}};$$

the proof was completed by Richmond & Wormald, again by showing that the symmetric polytopes were few enough to be ignored.

Of course, one can also try to count the combinatorial types of d-polytopes

Table B2. The number of d-polytopes with small numbers of vertices.

vertices	dimension d 3	4	5	6	7
4	1				
5	2	1			
6	7	4	1		
7	34	31	6	1	
8	257	1294	116	9	1
9	2606			379	12
10	32300				

for $d > 3$. The problem is of an essentially different nature to that of $d = 3$ (see page 290 of Grünbaum's book). However, substantial progress has been made recently, and it is now known that for each $d > 3$ there are constants a and b such that

$$v^{av} \leq \text{number of convex polytopes with } v \text{ vertices} \leq v^{bv}.$$

The left-hand inequality is due to Shemer; the right-hand one follows from the work of Goodman & Pollack and Alon; their exponent bv is a substantial improvement on $bv^{1/2d}$, which follows from Klee's upper bound theorem for simplicial polytopes. What is the correct constant in the exponent? The work cited gives $a = \frac{1}{2} + O(1)$ and $b = d^2[1 + O(1)]$ as $d \to \infty$. Reif has listed the 4-polytopes with 7 vertices, and Altshuler and Steinberg those with 8 vertices. There are $\lfloor \frac{1}{4}d^2 \rfloor$ d-polytopes with $d + 2$ vertices, and Lloyd has obtained a formula (occupying some five lines) that gives the exact number of combinatorial types of d-polytopes with $d + 3$ vertices. Some values are given in Table B2.

The d-polytopes of particular types (e.g., simplicial or neighborly) have been enumerated in certain cases (see Grünbaum or Altshuler & Steinberg).

N. Alon, The number of polytopes, configurations and real matroids, *Mathematika* **33** (1986) 62–71; *MR* **88c**:05038.

A. Altshuler, J. Bokowski & L. Steinberg, The classification of simplicial 3-spheres with nine vertices into polytopes and non-polytopes, *Discrete Math.* **31** (1980) 115–124; *MR* **81m**:52016.

A. Altshuler & L. Steinberg, Enumeration of the quasisimplicial 3-spheres and 4-polytopes with eight vertices, *Pacific J. Math.* **113** (1984) 269–288; *MR* **85h**:52006.

A. Altshuler & L. Steinberg, The complete enumeration of the 4-polytopes and 3-spheres with eight vertices, *Pacific J. Math.* **117** (1985) 1–16; *MR* **86f**:52009.

E. A. Bender, The number of 3-dimensional convex polyhedra, *Amer. Math. Monthly* **94** (1987) 7–21; *MR* **88b**:52010.

E. A. Bender & L. B. Richmond, The asymptotic enumeration of rooted convex polyhedra, *J. Combin. Theory Ser. B* **36** (1984) 276–283; *MR* **85m**:05004.

E. A. Bender & N. C. Wormald, Almost all convex polyhedra are asymmetric, *Canad. J. Math.* **37** (1985) 854–871; *MR* **87b**:52014.

D. Britton & J. D. Dunitz, A complete catalogue of polyhedra with eight or fewer vertices, *Acta Cryst. Sect. A* **29** (1973) 362–371.

A. Cayley, On the Δ-faced polygons in reference to the problem of the enumeration of polyhedra, *Mem. Lit. Philos. Soc. Manchester* **1** (1862) 248–256.

A. J. W Duijvestijn & P. J. Federico, The number of polyhedral (3-connected planar) graphs, *Math. Comput.* **37** (1981) 523–532; *MR* **83f**:05028.

P. Engel, On the enumeration of polyhedra, *Discrete Math.* **41** (1982) 215–218; *MR* **84h**:52001.

L. Euler, Elementa doctrinae solidorum, *Comment. Acad. Sci. Imp. Pertrop.* **4** (1752) 109–140.

P. J. Federico, The number of polyhedra, *Philips Res. Reps.* **30** (1975) 220–231.

P. J. Federico, Enumeration of polyhedra: the number of 9-hedra, *J. Combin. Theory* **7** (1969) 155–161; *MR* **39** #4746.

P. J. Federico, Polyhedra with 4 to 8 faces, *Geom. Dedicata* **3** (1974/75) 469–481; *MR* **51** #6585.

J. E. Goodman & R. Pollack, Upper bounds for configurations and polytopes in R^d, *Discrete Comput. Geom.* **1** (1986) 219–227; *MR* **87k**:52016.

J. E. Goodman & R. Pollack, There are asymptotically far fewer polytopes than we thought, *Bull. Amer. Math. Soc.* **14** (1986) 127–129; *MR* **87d**:52010.

D. W. Grace, Computer search for non-isomorphic convex polyhedra, Reprint CS15, Computer Sci. Dept., Stanford Univ., 1965.

B. Grünbaum, Polytopes, graphs and complexes, *Bull. Amer. Math. Soc.* **76** (1970) 1131–1201; *MR* **42** #959.

B. Grünbaum, [Gru].

O. Hermes, Die Formen der Vielflache, *J. Reine Angew Math.* **120** (1899) 27–59, 305–353; **122** (1900) 124–154; **123** (1901) 312–342.

T. P. Kirkman, On the representation and enumeration of polyhedra, *Mem. Lit. Philos. Soc. Manchester* (2) **12** (1854) 47–70.

V. Klee, On the number of vertices of a convex polytope, *Canad. J. Math.* **16** (1964) 701–720; *MR* **29** #3955.

V. A. Liskovets, Enumeration of nonisomorphic planar maps, *J. Graph Theory* **5** (1981) 115–117; *MR* **82d**:05067.

N. Linial, Hard enumeration problems in geometry and combinatorics, *SIAM J. Algebraic Discrete Methods* **7** (1986) 331–335; *MR* **87e**:68029.

E. K. Lloyd, The number of d-polytopes with $d + 3$ vertices, *Mathematika* **17** (1970) 120–132; *MR* **43** #1036.

M. Reif, 4-polytopes with 8 vertices, MSc thesis, Ben-Gurion University of the Negev.

L. B. Richmond & N. C. Wormald, The asymptotic number of convex polyhedra, *Trans. Amer. Math. Soc.* **273** (1982) 721–735; *MR* **84d**:52010.

I. Shemer, Neighborly polytopes, *Israel J. Math.* **43** (1982) 291–314; *MR* **84k**:52008.

J. Steiner, Problème de situation, *Ann. Math. Gergonne* **19** (1830) 36.

W. T. Tutte, A census of planar triangulations, *Canad. J. Math.* **14** (1962) 21–38; *MR* **24** #A695.

W. T. Tutte, A census of planar maps, *Canad. J. Math.* **15** (1963) 249–271; *MR* **26** #4343.

W. T. Tutte, On the enumeration of convex polyhedra, *J. Combin. Theory Ser. B* **28** (1980) 105–126; *MR* **81j**:05073.

W. T. Tutte, Convex polyhedra, in [DGS], 579–582; *MR* **84j**:52006.

B16. The sizes of the faces of a polyhedron.

B16. **The sizes of the faces of a polyhedron.** For which sequences $(p_3, p_4, p_5, \ldots)$ do there exist a polyhedron with p_3 triangular faces, p_4 quadrilateral faces, p_5 pentagonal faces, etc? This is an old and difficult problem. Using Euler's relation ($v - e + f = 2$, where v, e, and f are the numbers of vertices, edge, and faces) it follows easily that any sequence realizable by a convex polyhedron must satisfy

$$\sum_{n \geq 3} (6 - n)p_n \geq 12. \tag{1}$$

Note that the number of hexagons p_6, which plays a critical role in this work, is effectively omitted in this expression. In particular

$$\sum_{n \geq 3} (6 - n)p_n = 12 \tag{2}$$

if the sequence is realizable by a simple polyhedron (i.e., one with three edges adjacent to each vertex). Eberhard's theorem provides a partial converse of this, namely, that if $(p_3, p_4, p_5, p_7, p_8, \ldots)$ satisfies (1) then there exist values of p_6 for which the sequence is realizable by a simple polyhedron, and there is a similar result for 4-valent polyhedra. Another theorem of this nature, due to Grünbaum, concerns polyhedra with no 3- or 4-sided faces: if $p_6 \geq 8$ then the sequence $(0, 0, p_5, p_6, p_7, \ldots)$ is realized by a simple polyhedron, provided it satisfies (2). Roudneff has shown that

$$\sum_{n \geq 3} (n - 6)p_n \leq v - 12$$

for polyhedra with at least 3 faces of 6 or more sides, where v is the total number of vertices.

Find further inequalities that must be satisfied by the p_n and, ultimately, a set of inequalities characterizing the possible sequences.

One can also seek conditions for sequences to be realizable by polyhedra of certain types (e.g., Fisher considers the case of 5-valent polyhedra).

D. Barnette, On the p-vectors of 3-polytopes, *J. Combin. Theory* **7** (1969) 99–103; *MR* **39** #6165.

V. Eberhard, Zur Morphologie der Polyeder, Leipzig, 1981.

J. C. Fisher, An existence theorem for simple convex polyhedra, *Discrete Math.* **7** (1974) 75–87; *MR* **48** #12303.

J. C. Fisher, Five-valent convex polyhedra with prescribed faces, *J. Combin. Theory Ser. A* **18** (1975) 1–11; *MR* **51** #1612.

B. Grünbaum, [Gru], Sections 13.3 and 13.4.

B. Grünbaum, Problem 10, [Kle], 314–315.

B. Grünbaum, Some analogues of Eberhard's theorem on convex polytopes, *Israel J. Math.* **6** (1968) 398–411; *MR* **39** #6168.

B. Grünbaum, Polytopes, graphs and complexes, *Bull. Amer. Math. Soc.* **76** (1970) 1131–1201; *MR* **42** #959.

B. Grünbaum and J. Zaks, The existence of certain planar maps, *Discrete Math.* **10** (1974) 93–115; *MR* **50** #1949.

E. Jucovič, On the number of hexagons in a map, *J. Combin. Theory Ser. B* **10** (1971) 232–236; *MR* **43** #3910.

J. Malkevitch, Eberhard's theorem for 4-valent convex 3-polytopes, in [KB], 209–214.

J. P. Roudneff, An inequality for 3-polytopes, *J. Combin. Theory Ser. B* **42** (1987) 156–166; *MR* **88b**:52016.

B17. **Unimodality of *f*-vectors of polytopes.** The **f-vector** of a convex d-polytope P is defined by $(f_0, f_1, f_2, \ldots, f_{d-1})$, where P has f_0 vertices, f_1

edges, f_2 two-dimensional faces, ..., and f_{d-1} $(d - 1)$-dimensional faces. The unimodality conjecture, that for some k we have $f_0 \leq f_1 \leq \cdots \leq f_{k-1} \leq f_k \geq f_{k+1} \geq \cdots \geq f_{d-1}$, was apparently suggested by Motzkin and was published by Welsh.

Recently, Björner and Lee have shown that there are 24-dimensional simplicial polytopes with about 2.6×10^{11} vertices and $f_{14} > f_{15} < f_{16}$, and also 20-dimensional simplicial polytopes with 4.2×10^{12} vertices with $f_{11} > f_{12} < f_{13}$, thus disproving the conjecture. (A d-polytope is **simplicial** if all its $(d - 1)$-dimensional faces are $(d - 1)$-simplexes.) Are there smaller examples? In particular, what is the smallest dimension in which the unimodality conjecture fails, both in the simplicial case and in general? Björner claims that for simplicial d-polytopes, the conjecture is true if $d \leq 15$.

Some remarkable progress has been made recently on the characterization of the vectors that can occur as f-vectors of d-polytopes (see Lee for a survey of some of this work and many references). McMullen conjectured that a vector $(f_0, f_1, \ldots, f_{d-1})$ was the f-vector of a simplicial polytope if and only if the f_i satisfy a collection of rather complicated combinatorial equations and inequalities. The sufficiency of McMullen's conditions was established by Billera & Lee and the necessity by Stanley. Thus, at least in the simplicial case, the problem reduces to seeking non-unimodal solutions of a collection of inequalities. However, the form of the inequalities renders this far from trivial.

Björner points out that the counterexamples have implications for other "unimodality" conjectures that have been based on empirical evidence, such as the unimodality of the coefficients of the chromatic number of a graph. It may be that counterexamples are hard to find only because of their huge size.

L. J Billera & C. W. Lee, A proof of the sufficiency of McMullen's conditons for f-vectors for simplicial convex polytopes, *J. Combin. Theory Ser. A* **31** (1981) 237–255; *MR* **82m**:52006.

A. Björner, The unimodality conjecture for convex polytopes, *Bull. Amer. Math. Soc.* (*N.S.*) **4** (1981) 187–188; *MR* **82b**:52013.

C. W. Lee, Characterizing the numbers of faces of a simplicial convex polytope, in [KB], 21–38; *MR* **83h**:52008.

R. P. Stanley, The number of faces of a simplicial convex polytope, *Adv. Math.* **35** (1980) 236–238; *MR* **81f**:52014.

R. P. Stanley, The number of faces of simplicial polytopes and spheres, in [GLMP], 212–223.

D. J. A. Welsh, Problem 1, *Combinatorics*, D. J. A. Welsh & D. R. Woodall (eds.), Inst. Math. and Appl., Oxford, 1972, 357.

B18. **Inscribable and circumscribable polyhedra.** A convex polyhedron P is of **inscribable type** if there is a combinatorially equivalent polyhedron P' that can be inscribed in a sphere (i.e., with all the vertices of P' on the surface of the sphere). Similarly, P is of **circumscribable type** if an equivalent P' can be circumscribed about a sphere with every face touching the sphere. It follows

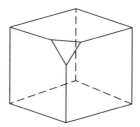

Figure B6.   An example of a noninscribable polyhedron—a truncated cube.

by polar reciprocity that a polyhedron is of inscribable type if and only if any dual polyhedron is of circumscribable type.

In 1832 Steiner asked whether every polyhedron is of inscribable type, and it was nearly a 100 years before Steinitz discovered examples (indeed infinite families of examples) that were not. The polyhedron of non-inscribable type with fewest faces is the cube truncated at one vertex (see Figure B6).

It would of course be nice to characterize the polyhedra of inscribable type, but as this may be over-optimistic, good necessary, or sufficient, conditions would be of interest. Do there exist polyhedra of non-inscribable type with all vertices having valence at least four?

Is every convex polyhedron equivalent to one with all its *edges* touching a sphere?

Schulte conjectures that, given any (3-dimensional) convex body K, there exists a convex polyhedron with no equivalent having all its faces tangent to K.

If P is any polyhedron and P^* is dual to P, can we always find P_1 equivalent to P and P_1^* equivalent to P^* such that the interior of each face of P_1 contains exactly one vertex of P_1^*?

Many further variants of these problems are given by Grünbaum & Shephard.

B. Grünbaum, [Gru].

B. Grünbaum, On Steinitz's theorem about non-inscribable polyhedra, *Ned. Akad. Wetenschap. Proc. A* **66** (1983) 452–455.

B. Grünbaum & E. Jucovič, On non-inscribable polytopes, *Czechoslovak Math. J.* **24** (1974) 424–429; *MR* **50** #1125.

B. Grünbaum & G. C. Shephard, Some problems on polyhedra, *J. Geom.* **29** (1987) 182–190; *MR* **88h**:52013.

E. Jucovič, On noninscribable polyhedra (Slovak), I, II, *Mat.-Fyz. Časopis Sloven. Akad. Vied* **15** (1965) 90–94; **16** (1966) 229–234; *MR* **31** #5153, **35** #868.

E. Schulte, Analogues of Steinitz's theorem about noninscribable polytopes, in [BF], 503–516; *MR* **89a**:52021.

S. Sever, On the non-inscribability of certain families of polyhedra, *Math. Slovava* **32** (1982) 23–34.

J. Steiner, Systematische Entwicklung der Abhängigkeit geometrischer Gestalten von einander, Fincke, Berlin, 1832.

E. Steinitz, Uber isoperimetrische Probleme bei konvexen Polyedern, *J. Reine Angew. Math.* **158** (1927) 129–153; **159** (1928) 133–143.

B19. **Truncating polyhedra.** The following beautiful conjecture is now known to be true—Hadwiger has shown it to be equivalent to the four color theorem! Let P be a convex polyhedron. Then there is a sequence of truncations of P such that all the faces of the resulting polyhedron have a multiple of three edges. (A polyhedron P is **truncated** to form a new polyhedron by slicing off a portion by a plane that separates one vertex of P from its remaining vertices.) Required: a simple proof!

K. Appel & W. Haken, Every plane map is four colorable, I, II, *Illinois J. Math.* **21** (1977) 429–567; *MR* **58** #27598.

H. Hadwiger, Ungelöste Probleme 17, *Elem. Math.* **12** (1957) 61–62.

G. C. Shephard, Twenty problems on convex polyhedra, I, II, *Math. Gaz.* **52** (1968) 136–147, 359–367; *MR* **37** #6833, **38** #5111.

T. L. Saaty & P. C. Kainen, *The Four-Color Problem*, Mc Graw-Hill, New York, 1977.

B20. **Lengths of paths on polyhedra.** The following is a problem from L. Moser's collection: What is the largest number $f(v)$ such that every convex polyhedron with v vertices has a simple (i.e., non-self-intersecting) path along its edges passing through $f(v)$ vertices? It is known from the work of Moon & Moser and Barnette that

$$2 \ln v/\ln 2 - 5 \leq f(v) < 9v^{\ln 2/\ln 3},$$

but what is the true order of magnitude? It is conjectured that $f(v) \leq v^{1/2}$. Many further results, where restrictions are placed on the number of sides of the faces and on the valence of the vertices, are given by Grünbaum & Walther in a paper that contains many further references and problems.

It seems appropriate to mention W.M. Hirsch's celebrated d-step conjecture at this juncture, though the outstanding problems concern higher-dimensional polytopes. While this conjecture is expressible in terms of edge-paths on polytopes, it is related to fundamental problems in linear programming and computational complexity. One form of the d-step conjecture is that if a (bounded) convex d-polytope P has f faces, then any pair of vertices of P may be joined by a path traversing at most $f - d$ edges. Klee & Walkup showed that if this could be proved for $f = 2d$ for all d, then the general case would follow. An equivalent conjecture is that any pair of vertices of P can be joined by an edge path that does not revisit any face of P. The conjecture is easy when $d = 2$ or 3 and has been proved by Klee for $d = 4$ and 5. The conjecture is open for $d = 6$, and the feeling is that it is false if d is sufficiently large. Of course, it is nevertheless desirable to obtain estimates for $\Delta(d, f)$, the maximum number of edges required to join a pair of vertices in any d-polytope with f faces. The best known estimates are due to Adler and Larman:

$$1 + \lfloor (f - d)(1 - 1/\lfloor 5d/4 \rfloor) \rfloor \leq \Delta(d, f) \leq 2^{d-3}f,$$

with

$$d \leq \Delta(d, 2d) \leq 2^{d-3}d$$

in the special case. Even if the d-step conjecture is false, it would be extremely useful to have good asymptotic estimates for $\Delta(d, f)$; in particular does $\Delta(d, 2d)$ increase linearly, quadratically, polynomially, or exponentially with d?

The problem of bounding path lengths in terms of the number of vertices is much simpler. It is well known that for a convex polyhedron with v vertices, every pair of vertices may be joined by a path of at most $\lfloor (v + 1)/3 \rfloor$ edges, and that this number cannot be reduced.

Further reading on topics of this type may be found in Chapters 16 and 17 (both written by Klee) of Grünbaum's book and in the survey by Klee & Kleinschmidt.

I. Adler, Lower bounds for maximum diameters of polytopes, *Math. Programming Study* **1** (1974) 11–19.

D. Barnette, Trees in polyhedral graphs, *Canad. J. Math.* **18** (1966) 731–736; *MR* **33** #3951.

B. Grünbaum, [Gru], Chapters 16 and 17.

B. Grünbaum & H. Walther, Shortness exponents of families of graphs, *J. Combin. Theory Ser. A* **14** (1973) 364–385; *MR* **47** #3242.

V. Klee, Diameter of polyhedral graphs, *Canad. J. Math.* **16** (1964) 602–614; *MR* **29** #2796.

V. Klee, Convex polyhedra and mathematical programming, Proc. Internat. Congress Math., Vancouver 1974, Canad. Math. Cong. 1975, 485–490; *MR* **55** #2127.

V. Klee, How many steps? in [KB], 1–6.

V. Klee & P. Kleinschmidt, The d-step conjecture and its relatives, *Math. Oper. Res.* **12** (1987) 718–755; *MR* **89a**:52018.

V. Klee & D. W. Walkup, The d-step conjecture for polyhedra of dimension $d < 6$, *Acta Math.* **117** (1967) 53–78; *MR* **34** #6639.

D. G. Larman, Paths on polytopes, *Proc. London Math. Soc.* (3) **20** (1970) 161–178; *MR* **40** #7942.

J. W. Moon & L. Moser, Simple paths on polyhedra, *Pacific J. Math.* **13** (1963) 629–631; *MR* **27** #4225.

B21. **Nets of polyhedra.** Consider a cardboard model of (the surface of) a polyhedron. A **net** of the polyhedron is formed by cutting along certain edges and unfolding the resulting connected set to lie flat. Nets of various polyhedra are shown in Figure B7 and many further examples are pictured in the book by Cundy & Rollett. Alexandrov's book discusses various mathematical aspects of nets.

For "well-known" polyhedra, the net never overlaps itself, but it is easy to find non-convex polyhedra for which it must always do so. For certain convex polyhedra (e.g., the snub dodecahedron) we can get an overlapping net if we cut perversely. Can we always find *some* way of cutting to get a non-overlapping net? If not, what is the smallest polyhedron with no such net? Does every convex polyhedron have a combinatorial equivalent with a non-overlapping net?

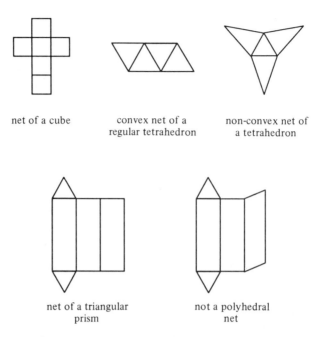

net of a cube   convex net of a   non-convex net of
                       regular tetrahedron   a tetrahedron

net of a triangular   not a polyhedral
        prism   net

Figure B7. A variety of nets of convex polyhedra.

Shephard considered a question of Sallee of when a polyhedron has a *convex* net [i.e., that is, when it can be unfolded to give a (non-overlapping) convex polygon]. This is very rarely possible, but if one considers which combinatorial types of polyhedra have a representative with a convex net, then the question becomes very interesting. Shephard lists several classes of polyhedra which have representatives with convex **Hamiltonian** nets (i.e., nets which consists of a strip of faces each adjacent to at most two others). He conjectures that any polyhedron that has a Hamiltonian path (i.e., one traversing each edge exactly once) is equivalent to one with a convex Hamiltonian net. This would imply that all polyhedra of less than 14 vertices are equivalent to ones with convex nets since such polyhedra all have a Hamiltonian path. Is there a 15 or 16 vertex polyhedron without a convex Hamiltonian net? Shephard gives an example with 17 vertices. He also conjectures that any polyhedron that has a combinatorial representative which unfolds to a *strictly* convex net must have a representative that unfolds to a convex Hamiltonian net. (A net is **strictly convex** if each edge comes from a single edge of the polyhedron.)

Surprisingly, it is possible to fold certain nets into polyhedra of differing combinatorial types by folding so that different combinations of edges are joined. Shephard asks if this can happen with a convex net. Is there a net that can be folded into k different combinatorial types of polyhedra, for each positive integer k?

It would be nice to have an algorithm to determine whether or not a given configuration of polygons is a net of some convex polyhedron. It would, for example, need to distinguish between the two configurations shown. Perhaps Alexandrov's "intrinsic metric" is relevant here.

A. D. Aleksandrov, *Konvex Polyeder*, Akademie Verlag, Berlin, 1958; *M R* **19**, 1192.
H. M. Cundy & A. P. Rollett, *Mathematical Models*, 2nd ed., Clarendon, Oxford, 1961; *M R* **23** #A1484.
G. C. Shephard, Convex polytopes with convex nets, *Math. Proc. Cambridge Philos. Soc.* **78** (1975) 389–403; *M R* **52** #11738.

B22. **Polyhedra with congruent faces.** For which triangles T does there exist a convex polyhedron with all of its faces congruent to T, either with or without reflections allowed? Certainly convex polyhedra can be constructed from any isosceles triangle, but which other triangles may be used? In each case, how many congruent triangles can be used, and how should they be arranged?

Similarly, for each n, which n-gons can be used to build up polyhedra with congruent faces?

B23. **Ordering the faces of a polyhedron.** The following question is due to Tverberg. Let $F_1, \ldots, F_n$ be the faces of a convex polyhedron. Is it always possible to choose the numbering of the faces so that for $i < n$ the face F_i abuts with at most four of the faces $F_1, \ldots, F_{i-1}$? This result would imply a weaker form of the four color theorem, in the case when one country is allowed to have the same color as some of its neighbors.

H. Tverberg, Problem 22, Durham Symposium, *Bull. London Math. Soc.* **8** (1976) 1–33.

B24. **The four color conjecture for toroidal polyhedra.** It is not hard to see that the four color theorem is equivalent to the statement that the faces of any convex polyhedron may be colored with four colors with faces with a common edge having different colors. It is also well-known that the surface of a smooth torus may be divided into regions that require seven colors. Thus, it was surprising when Barnette gave a simple argument to show that the faces of any toroidal polyhedron may be colored using only six colors. (**A toroidal**

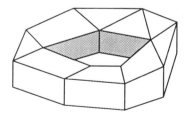

Figure B8.   A toroidal polyhedron.

polyhedron is made up by joining polygonal faces edge-to-edge so that the resulting figure is topologically equivalent to a torus, such as that shown in Figure B8. Of course, toroidal polyhedra are not convex.) However, Barnette conjectures the much stronger result: that the faces of every toroidal polyhedron can be colored using just four colors.

One can ask similar questions for the number of colors required to color the faces of non-convex polyhedra of other topological types. For which types does the polyhedron require fewer colors than for arbitrary maps?

D. W. Barnette, Coloring polyhedral manifolds, in [GLMP], 192–195; *MR* **86m**:05039.

B25. **Sequences of polygons and polyhedra.** Given an n-sided convex polygon P_1, define a sequence of n-gons $P_1, P_2, P_3, \ldots$ by taking the vertices of P_k to be the midpoints of the edges of P_{k-1}. It is clear that the polygons are decreasing in size and, since the P_k have a common vertex centroid (i.e., center of gravity of an equal mass at each vertex), the polygons converge to the vertex centroid of P_1. The interesting fact, however, is that the polygons become closer and closer in *shape* to an affine image of a regular n-gon (see Figure B9). More precisely, there are similarity transformations $\varphi_1, \varphi_2, \ldots$ such that the sequence $\varphi_1(P_1), \varphi_2(P_2), \ldots$ converges to an affine image of a regular n-gon. This beautiful result was first published by Darboux and has subsequently been rediscovered by several other authors.

One can define sequences of polyhedra in a way analogous to the above. For example, we can form sequences $P_1, P_2, \ldots$ where P_k is the polyhedron with vertices at the centroids of the faces of P_{k-1}; alternatively, we can use other types of face "centers" such as incenters, circumcenters, etc. Simple examples show that sequences of polyhedra constructed in this sort of way can behave in a very irregular and unstable manner. Can anything at all be said about the ultimate shape of such sequences of polyhedra?

The situation is rather different if the vertices of P_k are taken as the midpoints of the edges of P_{k-1}. Since P_k then contains the face-centroids of all its predecessors, the sequence of polygons converges to a nondegenerate convex body P, which Fejes Tóth calls the **kernel** of P_1. He conjectures that

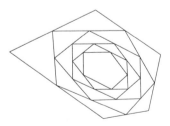

Figure B9.   A sequence of mid-point pentagons converging in shape to an affine regular pentagon.

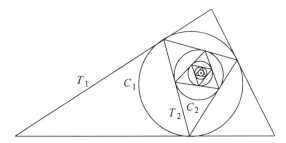

Figure B10.   An incircle–circumcircle sequence converging to the Poncelet point of T_1.

the ratio (volume of the kernel of P_1)/(volume of P_1) is minimum if P_1 is a tetrahedron. He also asks whether the kernel can ever be a *smooth* convex body—it need not have a twice differentiable surface, see Gardner & Kallay.

In the plane, Clarke asks whether the sequence of *pentagons* P_1, P_2, ..., where P_k is the pentagon formed by the five diagonals of P_{k-1}, has any interesting limiting properties. Is there a simple description of the limiting point in terms of the original pentagon?

Let T_1 be a given triangle, and form the sequence T_1, C_1, T_2, C_2, ..., where C_k is the incircle of T_k, and where the vertices of T_{k+1} are the points of contact of C_k with T_k, so that C_k is the circumcircle of T_{k+1} (see Figure B10). The triangles converge to a single point, called the **Poncelet** point x of T_1. Synge (see Moser & Pach) asks if x can be specified in finite terms (i.e., by a formula involving the vertices of T_1).

E. R. Berlekamp, E. N. Gilbert & F. W. Sinden, A polygon problem, *Amer. Math. Monthly* **72** (1965) 233–241; *MR* **31** #3925.

M. Bourdeau & S. Dubuc, L'itération de Fejes Tóth sur un polygone, *J. Geom.* **6** (1975) 65–75; *MR* **51** #6580.

G. Bruckner, Über n-Ecke, *Studia Sci. Math. Hungar.* **7** (1972) 429–435; *MR* **49** #1322.

J. H. Cadwell, A property of linear cyclic transformations, *Math. Gaz.* **37** (1953) 85–89; *MR* **15**, 55.

J. H. Cadwell, Nested polygons, *Topics in Recreational Mathematics*, Cambridge University Press, Cambridge, 1966, 22–31.

R. J. Clarke, Sequences of polygons, *Math. Mag.* **52** (1979) 102–105; *MR* **80c**:52004.

M. G. Darboux, *Bull. Sci. Math.* (2) 1878, 298–304.

P. J. Davis, Cyclic transformations of polygons and the generalized inverse, *Canad. J. Math.* **29** (1977) 756–770; *MR* **58** #22100.

P. J. Davis, *Circulant Matrices*, Wiley, New York, 1979; *MR* **81a**:15003.

G. Fejes Tóth, Iteration processes leading to a regular polygon (Hungarian), *Mat. Lapok* **23** (1972) 135–141; *MR* **49** #3683.

L. Fejes Tóth, Iteration methods for convex polygons (Hungarian), *Mat. Lapok* **20** (1969) 15–23; *MR* **40** #7940.

L Fejes Tóth, Sequences of polyhedra, *Amer. Math. Monthly* **88** (1981) 145–146.

R. J. Gardner & M. Kallay, Subdivision algorithms and the kernel of a polyhedron, to appear.

J. C. Fisher, D. Ruoff & J. Shilleto, Polygons and polynomials, in [DGS], 321–333.

J. C. Fisher, D. Ruoff & J. Shilleto, Perpendicular polygons, *Amer. Math. Monthly* **92** (1985) 23–37.

F. Kárteszi, A pair of affinely regular pentagons generated from a convex pentagon (Hungarian), *Mat. Lapok* 20 (1969) 7–13; *MR* **40** #7941.

G. Korchmáros, An iteration process leading to affinely regular polygons (Hungarian), *Mat. Lapok* **20** (1969) 405–411; *MR* **42** #6717.

G. Lükő, Certain sequences of inscribed polygons, *Period. Math. Hungar.* **3** (1973) 255–260; *MR* **48** #12291.

W. Moser & J. Pach, Problem 69, [MP].

B. H. Neumann, Some remarks on polygons, *J. London Math. Soc.* **16** (1941) 230–245; *MR* **4**, 51.

H. Reichardt, Bestätigung einer Vermutung von Fejes Tóth, *Rev. Roumaine Math. Pure Appl.* **15** (1970) 1513–1518; *MR* **43** #1037.

M. Ruda, On a conjecture of Lászlo Fejes Tóth (Hungarian), *Magyar Tud. Akad. Mat. Fiz. Oszt. Közl.* **19** (1970) 375–381; *MR* **41** #4382.

M. Ruda, A theorem on the bisection of polygons (Hungarian), *Magyar Tud. Akad. Mat. Fiz. Oszt. Közl.* **22** (1974) 201–213; *MR* **53** #9037.

D. B. Shapiro, A periodicity problem in plane geometry, *Amer. Math. Monthly* **99** (1984) 97–108.

A. Szép, Linear iterations of polygons (Hungarian), *Mat. Lapok* **21** (1970) 255–260.

E. T. H. Wong, Polygons, circulant matrices and Moore–Penrose inverses, *Amer. Math. Monthly* **88** (1981) 509–515; *MR* **82m**:15004.

C. Tiling and Dissection

The art of tiling goes back thousands of years. For centuries builders of floors and walls have sought out arrangements that are pleasing to the eye, and many ancient tilings and mosaics survive today. More recently, similar regular arrangements have been observed in the molecular structure of crystals. Repeating patterns have also been used in modern drawings (e.g., those by Escher). Tilings frequently occur in books on recreational mathematics; indeed, amateur mathematicians have contributed much to this subject.

By a tiling of the plane (or, more generally, $\mathbb{R}^d$) we mean an exact covering of the plane (or $\mathbb{R}^d$) by a collection of sets without gaps or overlaps except perhaps at the boundaries of the sets. More precisely, a collection of sets (the "**tiles**") $\{T_1, T_2, \dots\}$ are a **tiling** or **tesselation** of the plane (or $\mathbb{R}^d$) if their union is the whole plane (or the whole of $\mathbb{R}^d$), but the interiors of different tiles are disjoint. Similarly, a tiling of an arbitrary set E is a collection $\{T_1, T_2, \dots\}$ with disjoint interiors and with union E. Sometimes it is more natural to think of E as being cut up into (essentially) disjoint pieces; thus we may refer to $\{T_1, T_2, \dots\}$ as a **dissection** of E.

Usually, the tiles T_i are closed and bounded, and frequently they are polygons. Very often, the tiles are all congruent to one another, or at least congruent to one of a small number of **prototiles**. Typically, one is interested in questions such as which polygons can be used as prototiles in a tiling of the plane. An enormous amount of deep research has gone into classifying tilings of particular types.

Very occasionally we require tilings or dissections of E to be **topologically exact**, in the sense that the tiles are completely disjoint though covering the whole of E. Thus, while the closed squares $\{[n, n + 1] \times [m, m + 1] : n, m \in \mathbb{Z}\}$ would normally be regarded as tiling the plane, the half-open squares $\{[n, n + 1) \times [m, m + 1) : n, m \in \mathbb{Z}\}$ provide a tiling that is topologically exact.

The book by Grünbaum & Shephard is the definitive work on tiling. It contains examples, diagrams, and tabulations of many types of tiling, as well as a full bibliography and further problems. It is indispensible to the serious student of the subject.

B. Grünbaum & G. C. Shephard, [GS].

C1. Conway's fried potato problem.

In order to fry it as expeditiously as possible, Conway wishes to slice a given convex potato into n pieces by $n - 1$ successive plane cuts (just one piece being divided by each cut), so as to minimize the greatest inradius of the pieces, which we denote by r.

For example, if $n = 2$ and the potato is a regular tetrahedron of edge length one, then there are at least three "obvious" ways in which to slice it (see Figure C1):

(a) through an edge and the midpoint of the opposite edge; $r = \frac{1}{5}(\sqrt{6} - 1) = 0.2899...$
(b) parallel to a face at one-third of its height; $r = \frac{1}{9}\sqrt{2} = 0.2722...$
(c) parallel to a pair of opposite edges; $r = \frac{1}{4}(\sqrt{6} - \sqrt{2}) = 0.2588....$

Thus (a) and (b) are about 12% and 5% worse than (c). Is (c) the best way to do it?

In general, does the best way always involve $n - 1$ parallel cuts, equally spaced between planes of support which define the (minimum) width of the potato?

There are similar slicing problems to minimize other measures of the chips, such as diameter or circumradius. If the potato is to be sliced so that all the chips pass through a circular hole of smallest possible diameter, must all cuts be perpendicular to some plane?

A very different set of problems arises if pieces are held together after each cut, so that up to $\binom{n+1}{3} + \binom{n+1}{1}$ pieces are formed by n cuts. The problems become even more complicated if the pieces can be rearranged at each stage before a plane cuts through them all, yielding as many as 2^n pieces.

There seems to be some relation between this and a problem of Hajós. If T is a tetrahedron, with $a(T)$ its longest edge, and $r(T)$ the radius of the

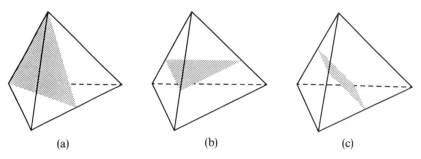

(a)   (b)   (c)

Figure C1.   Three ways of slicing a tetrahedron.

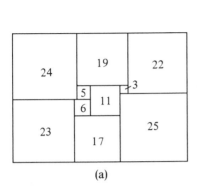 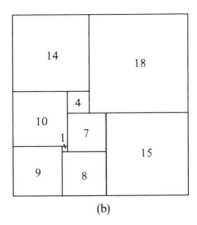

(a) (b)

Figure C2. Squared rectangles due to Moroń: (a) of order 10 and (b) of order 9, the smallest order possible.

inscribed sphere, give an effective construction for decomposing an arbitrary T into a finite number of tetrahedra, and then for each of these, and so on, so that none of the resulting tetrahedra are too long and thin [i.e., so that $a(T)/r(T)$ remains bounded above].

G. Hajós, Problem 122, *Colloq. Math.* **3** (1954) 74.

C2. Squaring the square. The problem of dissecting a square into smaller squares, all of different sizes, has fascinated many people. Although the main problems have been settled, the literature is scattered, so we collect it here.

The sides of the smaller squares have to be rational multiples of the original square side, so we lose nothing by working in integers. For many years the construction was thought to be impossible, though increasing numbers of squared rectangles were produced.

The **order** of a squared rectangle is the number of squares into which it is dissected. The rectangle is **perfect** if no two squares are equal; **compound** if it contains a squared (sub)rectangle, and **simple** if it is not compound.

The first perfect squared rectangle [see Figure C2(a)] was published in 1925 by Moroń: it is simple, and of order 10. He also gave an order 9 rectangle [see Figure C2(b)]; this is the smallest order possible (see Rouse Ball).

In 1939 Sprague produced a compound squared square of order 55 and a year later the classical paper of Brooks, Smith, Stone & Tutte appeared, relating the problem to flows in networks. Willcocks constructed a simple squared square of order 24 [see Figure C3(b)] which, for many years, was believed to be the least possible order.

The advent of computers allowed extensive searches to be made, notably by Bouwkamp, and later by Duijvestijn, culminating in 1978 with the discovery of the unique simple perfect squared square of lowest possible order, 21 (see Figure C3(a)).

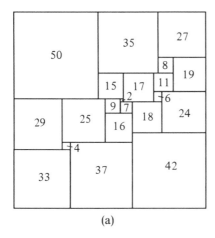
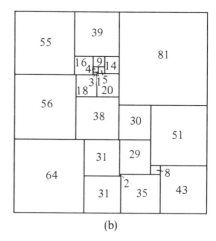

(a)   (b)

Figure C3. Squared squares: (a) of order 21, the least order possible, due to Duijvestijn, and (b) of order 24 due to Willcocks.

Moroń has asked if, in a squared rectangle, there are always at least two squares which are surrounded by (four) larger ones. Also if, in a squared rectangle whose sides are relatively prime, there is always a square whose side is a perfect square. Probably not. And is there a rectangle which can be squared in two distinct ways with no square occurring in both squarings? Probably so.

C. J. Bouwkamp, On the dissection of rectangles into squares I, *Proc. Nederl. Akad. Wetensch.* **49** (1946) 1176–1188 = *Indagationes Math.* **8** (1946) 724–736; II, III ibid. **50** (1947) 58–71, 72–78 = **9** (1947) 43–46, 57–63; *MR* **8**, 398.

C. J. Bouwkamp, On the construction of simple perfect squared squares, *ibid.* **50** (1947) 1296–1299 = **9** (1947) 622–625; *MR* **9**, 332.

C. J. Bouwkamp, On some special squared rectangles, *J. Combin. Theory Ser. B* **10** (1971) 206–211; *MR* **43** #79.

C. J. Bouwkamp, A. J. W. Duijvestijn & P. Medema, Catalogue of simple squared rectangles of orders nine through fourteen and their elements, Dept. Math. Tech. Hogeschool, Eindhoven, 1960; *MR* **23** #A1554.

C. J. Bouwkamp, A. J. W. Duijvestijn & P. Medema, Tables relating to simple squared rectangles of orders nine through fifteen, *ibid.* 1960; *MR* **23** #A1555.

R. L. Brooks, A procedure for dissecting a rectangle into squares, and an example for the rectangle whose sides are in the ratio 2 : 1, *J. Combin. Theory* **8** (1970) 232–243; *MR* **40** #7136.

R. L. Brooks, C. A. B. Smith, A. H. Stone & W. T. Tutte, The dissection of rectangles into squares, *Duke Math. J.* **7** (1940) 312–340; *MR* **2**, 153.

R. L. Brooks, C. A. B. Smith, A. H. Stone & W. T. Tutte, A simple perfect square, *Proc. Nederl. Akad. Wetensch.* **50** (1947) 1300–1301 = *Indagationes Math.* **9** (1947) 626–627; *MR* **9**, 332.

A. J. W. Duijvestijn, Electronic computation of squared rectangles, thesis, Tech. Hogeschool, Eindhoven, 1962; *MR* **26** #2036.

A. J. W. Duijvestijn, An algorithm that investigates the planarity of a network, *Computers in Mathematical Research*, North-Holland, Amsterdam, 1968, 68–83; *MR* **39** #5248.

A. J. W. Duijvestijn, Simple perfect squared square of lowest order, *J. Combin. Theory Ser. B* **25** (1978) 240–243; *M R* **80a**:05051.

A. J. W. Duijvestijn, P. J. Federico & P. Leeuw, Compound perfect squares, *Amer. Math. Monthly* **89** (1982) 15–32; *M R* **83h**:05031.

P. J. Federico, Note on some low-order perfect squared squares, *Canad. J. Math.* **15** (1963) 350–362; *M R* **26** #2926.

P. J. Federico, A Fibonacci perfect squared square, *Amer. Math. Monthly* **71** (1964) 404–406; *M R* **29** #5158.

P. J. Federico, Some simple perfect 2 × 1 rectangles, *J. Combin. Theory* **8** (1970) 244–246; *M R* **40** #7137.

P. J. Federico, Squaring rectangles and squares; a historical review with annotated bibliography, in *Graph Theory and Related Topics* J. A. Bondy and U. S. R. Murty (eds.), Academic Press, New York, 1979, 173–196; *M R* **80h**:05019.

M. Gardner, Mathematical games: squaring the square, *Scientific Amer.* **199**, No. 5 (Nov. 1958) 136–142.

M. Gardner, *Second Scientific American Book of Mathematical Puzzles and Diversions*, W. H. Freeman, San Francisco, 1961, 136–142.

N. D. Kazarinoff & R. Weitzenkamp, Squaring rectangles and squares, *Amer. Math. Monthly* **80** (1973) 877–888; *M R* **49** #10604.

N. D. Kazarinoff & R. Weitzenkamp, On the existence of compound perfect squared squares of small order, *J. Combin. Theory Ser. B* **14** (1973) 163–179; *M R* **47** #3191.

R. D. Mauldin, [Mau], Problem 59.

H. Meschkowski, [Mes], Chapter VII.

Z. Moroń, On the dissection of rectangles into squares, *Wiadom Mat.* (2) **1** (1955) 75–94; *M R* **16**, 1046.

Z. Moroń, On almost-perfect decompositions of rectangles (Polish), *Wiadom Mat.* (2) **1** (1955/56) 175–179; *M R* **22** #2573.

Z. Moroń, Problems 66–68, *Colloq. Math.* **2** (1951) 60–61.

W. W. Rouse Ball, *Mathematical Recreations and Essays*, 10th ed., Macmillan, New York, 1931, 93; 12th ed., University of Toronto Press, Toronto, 1974, 115.

R. P. Sprague, Beispiel einer Zerlegung des Quadrats in lauter vershiedene quadrate, *Math. Z.* **45** (1939) 607–608.

R. P. Sprague, Über die Zerlegung von Rechtecken in lauter verschiedene Quadrate, *J. Reine Angew. Math.* **182** (1940) 60–64; *M R* **2**, 11.

R. P. Sprague, Zur abschätzung der Mindestzahl inkongruenter Quadrate, die ein gegebenes Rechteck ausfüllen, *Math. Z.* **46** (1940) 460–471.

W. T. Tutte, A note to a paper by C. J. Bouwkamp, *Proc. Nederl. Akad. Wetensch.* **51** (1948) 280–282 = *Indagationes Math.* **10** (1948) 106–108; *M R* **9**, 571.

W. T. Tutte, The dissection of equilateral triangles into equilateral triangles, *Proc. Cambridge Philos. Soc.* **44** (1948) 463–482; *M R* **10**, 319.

W. T. Tutte, Squaring the square, *Canad. J. Math.* **2** (1950) 197–209; *M R* **12**, 118.

W. T. Tutte, The quest of the perfect square, *Amer. Math. Monthly* **72** (1965) 29–35; *M R* **32** #2338.

W. T. Tutte, Squared rectangles, *Proc. IBM Sci. Comput. Sympos. Combin. Problems*, Yorktown Heights New York, 1964, 3–9; *M R* **35** #5346.

T. H. Willcocks, Fairy Chess Review 7 (Aug/Oct. 1948).

T. H. Willcocks, A note on some perfect squared squares, *Canad. J. Math.* **3** (1951) 304–308; *M R* **13**, 264.

T. H. Willcocks, Some squared squares and rectangles, *J. Combin. Theory* **3** (1967) 54–56; *M R* **35** #5347.

C3. Mrs. Perkins's quilt.

Here we are concerned with imperfect squarings of the square. Let $f(n)$ be the smallest number of squares, greater than one,

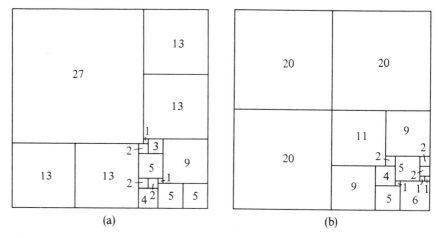

Figure C4. Two (imperfect) tilings of a square of side 40 using 16 integer sided tiles, the least number possible.

into which one can cut an n by n square, cutting only along the lines of the lattice. We will demand that the dissections are prime (i.e., that the greatest common factor of the square sides is one, so that the dissection is not a scalar multiple of another solution). On the other hand, we do allow **nonprimitive** solutions which contain subdivisions that are scalar multiples of smaller solutions. For example, Figure C4 shows that $f(40) = 16$; the second realization is nonprimitive as it contains a twenty-fold enlarged copy of the picture of $f(2) = 4$. It is known that $\log_2 n < f(n) < 6 \log_2 n$. The lower bound is due to Conway and the upper bound to Trustrum. The actual values of $f(n)$ are known, with some slightly increasing lack of confidence, for $n \leq 100$:

$f(n) =$	1	4	6	7	8	9	10	11	12	13	14
$n =$	1	2	3	4	5	6,7	8,9	10–13	14–17	18–23	24–29

$f(n) =$	15	16	17	18	19
$n =$	30–39, 41	40, 42–50	51–66, 68, 70	67, 69, 71–87	88–100

In the classical paper of Brooks, Smith, Stone & Tutte (see Section C2) it is shown that a given rectangle can be dissected into n squares in only a finite number of ways. So Selfridge observes that there is a prime $f = f(n)$ such that the $f \times f$ squares cannot be dissected into n smaller squares. In fact all but a finite number of primes have this property. For each n, what is the related set of primes?

J. H. Conway, Mrs. Perkins's quilt, *Proc. Cambridge Philos. Soc.* **60** (1964) 363–368; *MR* **29** #4698.

H. E. Dudeney, Mrs. Perkins' quilt, Problem 173, *Amusements in Mathematics*, Thomas Nelson and Sons, 1917.

H. E. Dudeney, Problem 177, *Puzzles and Curious Problems*, Thomas Nelson and Sons, 1931.

M. Gardner, *Mathematical Carnival*, Alfred A Knopf, New York; Random House, Toronto, 1965, Chap. 11.

S. Loyd, *Cyclopedia of Puzzles*, Lamb Publishing Co., 1914, 39,65.

G. B. Trustrum, Mrs. Perkins's quilt, *Proc. Cambridge Philos. Soc.* **61** (1965) 7–11; *MR* **30** # 1066.

C4. Decomposing a square or a cube into n smaller ones.

The following problem has been attributed to Hadwiger though it was also proposed by Fine & Niven. Let $D(d)$ be the set of integers n such that there is a decomposition of the d-dimensional cube into n homothetic d-cubes of smaller, not necessarily equal size. For example, $D(2)$ consists of all positive integers except 2, 3, and 5 (see Figure C5). What is $D(d)$ for $d \geq 3$? Let $c(d)$ be the smallest number such that if $n \geq c(d)$, then n belongs to $D(d)$ [e.g., $c(2) = 6$ and probably $c(3) = 48$]. The decomposition of the 3-cube into 54 smaller cubes is elusive, but was reconstructed by Doris Rychener and by A. Zbinden, noting that $8^3 = 6 \times 4^3 + 2 \times 3^3 + 4 \times 2^3 + 42 \times 1^3$. According to William Scott, $c(4) \leq 854$ and $c(5) \leq 1891$. More generally, Meier proved that $c(d) \leq (2^d - 2)[(2^d - 1)^d - (2^d - 2)^d - 1] + 1$ and Erdős used a theorem of A. Bauer to improve this bound to $(e - 1)(2d)^d$. But the actual value is probably much smaller for large d.

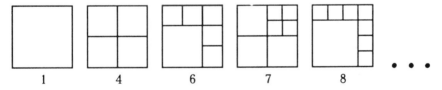

| 1 | 4 | 6 | 7 | 8 | · · · |

Figure C5. A square may be divided into n smaller squares for all positive integers n except 2, 3, and 5.

P. Erdős, Remarks on some problems in number theory, *Proc. Conf. of Balkan Math., Belgrade*, June 1974.

N. J. Fine & I. Niven, Problem E724, *Amer. Math. Monthly* **53** (1946) 271; solution **54** (1947) 41–42.

R. K. Guy, Monthly research problems, *Amer. Math. Monthly* **82** (1975) 1001.

R. K. Guy, Monthly research problems, *Amer. Math. Monthly* **84** (1977) 809–810.

C. Meier, Decomposition of a cube into smaller cubes, *Amer. Math. Monthly* **81** (1974) 630–632.

C5. Tiling with incomparable rectangles and cuboids.

Otokar Grošek asked how few incomparable cuboids will tile the cube. He asked the question in any number of dimensions, but we will start in two.

Two rectangles are **incomparable** if neither will fit inside the other with sides parallel, that is, if their dimensions can be written $a_1 \times a_2$ and $b_1 \times b_2$ with $a_1 < b_1 \leq b_2 < a_2$. The problem of tiling a rectangle with at least two incomparable rectangles was posed by Reingold and solved by several people who showed that no rectangle can be incomparably tiled with six or fewer tiles.

If we confine our attention to integer measurements we lose very little generality, and we can ask questions of the size of rectangles that permit tilings. Yao, Reingold & Sands show that the 13×22 rectangle of Figure C6(a) is the smallest integer rectangle that can be tiled with 7 incomparable rectangles, whether size is measured by area or by perimeter. The 34×34 square of Figure C6(b) is the smallest integer square which can be so tiled, though Figure C6(c) shows a smaller 27×27 square using 8 such tiles; it is not known whether this is the smallest square possible for any number of incomparable rectangles. In fact for each $k \geq 7$, there are only finitely many integers n such that the $n \times n$ square cannot not be tiled by k incomparable rectangles. For tilings with large numbers of pieces, are there incomparable tilings with the tiles arbitrarily close to squares (i.e., with a_1/a_2 etc., arbitrarily close to 1)?

In higher dimensions, if two cuboids are comparable, then they may be placed alongside each other with corresponding dimensions in increasing order, so two d-dimensional cuboids are incomparable if their dimensions, written in increasing order $a_1 \leq a_2 \leq \cdots \leq a_d, b_1 \leq b_2 \leq \cdots \leq b_d$, are such that there is an i and a j with $a_i < b_i$ and $a_j > b_j$. Trivial incomparable tilings of higher-dimensional cuboids can be obtained by erecting prisms with a lower-dimensional incomparable tiling as base. For example, cuboids of height

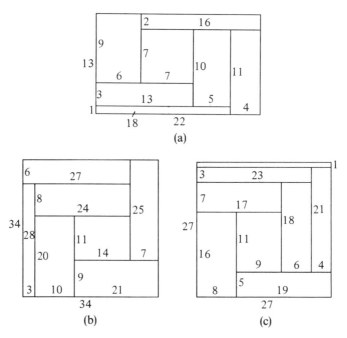

Figure C6.   Tilings with incomparable rectangles. (a) The smallest rectangle that can be tiled. (b) The smallest square that can be tiled with seven rectangles. (c) A smaller square that can be tiled with eight rectangles.

1, 34, or 27, erected on Figures C6(a) to C6(c), give what might be thought to be optimal tilings of a cuboid or cube in three dimensions. But the first of these is optimal neither by sense of volume, nor by number of pieces. Jepsen tiled a $3 \times 5 \times 9$ cuboid with just the 6 incomparable pieces $1 \times 1 \times 9$, $1 \times 2 \times 7$, $1 \times 4 \times 6$, $2 \times 2 \times 5$, $2 \times 4 \times 4$, and $3 \times 3 \times 4$ and showed that this is the smallest cuboid that can be tiled.

Jepsen also showed that there is no 6-piece incomparable tiling of the cube, but gave a non-trivial (not all pieces have the cube edge as a dimension) 7-piece incomparable tiling of the $10 \times 10 \times 10$ cube. Is this the smallest cube that can be so tiled?

Sands's tiling implies a trivial 6-piece incomparable tiling of the 4-dimensional cuboid. What is the optimal non-trivial incomparable tiling of a 4-dimensional cuboid (a) in terms of volume and (b) in terms of number of pieces?

O. Grošek, How few incomparable cuboids will tile the cube? *Amer. Math. Monthly* **91** (1984) 624–629.

C. M. Jepsen, Tiling with incomparable cuboids, *Math. Mag.* **59** (1986) 283–292; *MR* **89a**:05051.

A. C. C. Yao, E. M. Reingold & B. Sands, Tiling with incomparable rectangles, *J. Recreational Math.* **8** (1975/76) 112–119; *MR* **56** #5416.

C6. Cutting up squares, circles and polygons.
If a square, or indeed any rectangle, is cut (by rectifiable Jordan arcs) into 3, 5, 7, or 9 congruent parts (i.e., with congruent closure) are the parts necessarily rectangles? For even numbers, this is trivially false (see Figure C7). For larger odd numbers non-rectangular dissections exist. Klarner, in the article where the problem first appears in print, dissects a rectangle into 15 congruent trominoes and one into 11 congruent hexominoes (see Figure C7).

Another conjecture is that if a square is cut into an odd number of congruent convex sets, then they must be rectangles. In this direction Monsky used valuation theory to show that a rectangle cannot be cut up into an odd number of triangles of equal area. A direct proof of this fact would be of interest. Moreover, Hales & Straus showed that, in general, quadrilaterals cannot be dissected into triangles of equal areas. Further, Kasimatis proved that a regular n-gon ($n \geq 5$) can be partitioned into m triangles of equal area if and only if m is a multiple of n.

A problem with a number theoretic flavor, credited to Ihringer by Moser & Pach, is whether there are just finitely many ways to dissect a square into n rectangles of equal area. If so, estimate the number of such dissections for each n. Scheithaver & Terno show that the number must increase at least exponentially with n.

Stein asks whether it is possible to partition the unit circle into congruent pieces so that the center is in the interior of one of the pieces? It need not be on the boundary of all the pieces, as Figure C8 shows. Is it true that the pieces must have a diameter of at least one? This first question is also of interest for the regular n-gon, $n \geq 5$.

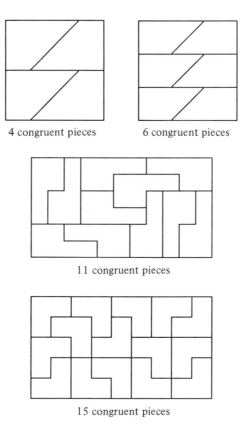

4 congruent pieces   6 congruent pieces

11 congruent pieces

15 congruent pieces

Figure C7.   Nontrivial division of a rectangle into congruent pieces.

It is easy to invent other dissection problems of this type in two or more dimensions. Danzer asks whether every convex polyhedron P admits a dissection into polyhedra all of which are combinatorially equivalent to P (see Chapter B, Introduction). A wide ranging conjecture of McMullen is that if K is a convex body in $\mathbb{R}^d$ ($d \geq 3$) which may be dissected into several congruent pieces, each similar to K, then K is a polytope combinatorially equivalent to a Cartesian product of simplexes. In a similar vein, Valette asks whether a convex body K in $\mathbb{R}^d$ ($d \geq 2$) that admits a dissection into several bodies each affinely equivalent to K, is either a polytope or an affine copy of a product of a convex body and a polytope with more than one point.

One can also consider the "topological" version of these questions. Can we dissect a (topologically closed or open) square, disk, equilateral triangle, etc. into $n > 2$ pieces that are exactly congruent? (The obvious dissections are not topologically congruent at the boundary.) For $n = 2$ the impossibility for any strictly convex body was proved by Puppe and others (see van der Waerden). Wagon shows that the ball in $\mathbb{R}^d$ cannot be dissected into n topologically congruent pieces if $2 \leq n \leq d$, but leaves open the case where $n > d$. What about topologically congruent dissections into countably many parts?

Figure C8.   Division of a circle into congruent pieces.

L. Danzer, Problem 9, in Durham Symposium on the relations between infinite-dimensional and finite-dimensional convexity, *Bull. London Math. Soc.* **8** (1976) 1–33.

M. Dehn, Zerlegung von Rechtecke, *Math. Ann.* **57** (1903) 314–332.

A. W. Hales & E. G. Straus, Projective colorings, *Pacific J. Math.* **99** (1982) 31–43; *MR* **83h**:51005.

E. A. Kasimatis, Dissections of regular polygons into triangles of equal areas, *Discrete Comput. Geom.* **4** (1989) 375–381.

E. A. Kasimatis & S. K. Stein, Equidissections of polygons, *Discrete Math.* **85** (1990) 281–294.

D. A. Klarner, Packing a rectangle with congruent *n*-ominoes, *J. Combin. Theory* **7** (1969) 107–115; *MR* **40** #1894.

H. Lindgren, *Geometric Dissections*, Van Nostrand, Princeton, 1964.

P. McMullen, Problem 60, [TW], 269.

P. Monsky, On dividing a square into triangles, *Amer. Math. Monthly* **77** (1970) 161–164; *MR* **40** #5454.

W. Moser & J. Pach, Problem 62, [MP].

S. K. Murrary, Problem 85.10, *Math. Intelligencer* **7** no. 4 (1985) 39.

G. Scheithauer & J. Terno, The partition of a square in rectangles of equal areas, *J. Inform. Process. Cybernet.* **24** (1988) 189–200; *MR* **90a**:52030.

J. S. Thomas, A dissection problem, *Math. Mag.* **41** (1968) 187–191; *MR* **38** #5099.

G. Valette, Problem 85, [TW], 274.

B. L. van der Waerden, Aufgabe Nr 51, *Elem. Math.* **4** (1949) **18**, 140.

S. Wagon, Partitioning intervals, spheres and balls into congruent pieces, *Canad. Math. Bull.* **26** (1983) 337–340; *MR* **84a**:52025.

C7. Dissecting a polygon into nearly equilateral triangles.

Every polygon P (convex or re-entrant) can be dissected into acute-angled triangles (see Manheimer). On the other hand P can be dissected into triangles with all angles less than $\frac{\pi}{3} + \varepsilon$ for all $\varepsilon > 0$ if and only if all the angles of P are multiples of $\frac{\pi}{3}$.

Let $\theta(P)$ be the infimum value of θ such that P can be divided into triangles with all angles at most θ. Does $\theta(P)$ depend only on the values of the interior angles $\theta_1, \ldots, \theta_n$ of P? If so, find a formula for $\theta(P)$ in terms of the θ_i. Gerver proves that if $\theta_i \geq \pi/5$ $(1 \leq i \leq n)$ then $\theta(P) \leq 2\pi/5$.

Denote by C_α the condition:

$$\text{the set } \{\theta_1, \ldots, \theta_n\} \text{ can be partitioned into sets} \tag{C_α}$$
$$R_1, R_2, \ldots \text{ with } k(\pi - 2\alpha) \leq \theta_i \leq k\alpha \text{ if } \theta_i \in R_k.$$

Gerver uses Euler's formula to show that

(i) $\theta(P) \le \alpha$ for $\alpha \ge 2\pi/5$ if (C_α) holds.

(ii) $\theta(P) \le \alpha$ for $\alpha \ge 5\pi/14$ if (C_α) holds with the R_k satisfying $\sum_k (3 - k)|R_k| \ge 6$.

(iii) $\theta(P) \le \alpha$ for $\alpha \ge \pi/3$ if (C_α) holds with the R_k satisfing $\sum_k (3 - k)|R_k| = 6$.

He conjectures that these conditions are also sufficient for $\theta(P) \le \alpha$, and also that, if $\theta(P) > \pi/3$, then there is actually a finite dissection into triangles with the greatest angle equal to $\theta(P)$.

One can investigate how many triangles are needed for a dissection with all the angles "small." For example, a square can be divided into r acute-angled triangles if $r = 8$ or $r \ge 10$ (see Cassidy & Lord).

B. S. Baker, E. Grosse & C. S. Rafferty, Non-obtuse triangulations of polygons, *Discrete Comput. Geom.* **3** (1988) 147–168.

C. Cassidy & G. Lord, A square acutely triangulated, *J. Recreational Math.* **13** (1980/81) 263–268; *MR* **84j**:51036.

M. Gardner, Mathematical games, *Scientific American* **202** (March 1960) 173–178.

M. R. Garey, D. S. Johnson, F. R. Preparata & R. E. Tarjan, Triangulating a simple polygon, *Inform. Process. Lett.* **7** (1978) 175–179; *MR* **58** #271.

J. L. Gerver, The dissection of a polygon into nearly equilateral triangles, *Geom. Dedicata* **16** (1984) 93–106; *MR* **85h**:51038.

W. Manheimer, Solution to Problem E1406, *Amer. Math. Monthly* **67** (1960) 177–178.

R. Sibson, Locally equiangular triangulations, *Comput. J.* **21** (1978) 243–245; *MR* **80d**:52018.

C8. Dissecting the sphere into small congruent pieces.

The following is a "Scottish Book" problem due to Ruziewicz: Given $\varepsilon > 0$, can we partition the surface of the unit sphere into a finite number of congruent (connected) pieces of diameter less than ε? The problem may have a different nature according to whether the boundaries are (a) spherical polygons, (b) curves of finite length, or (c) sets of area zero. If this is impossible, it would be interesting to find the smallest possible value of ε.

The spherical icosidodecahedron provides a nontrivial decomposition of the sphere into 120 congruent triangles (see Wenninger). Can this number be increased without allowing the triangles to become "too thin," say with each side less than $\frac{1}{3}\pi$? (Without some such condition trivial dissections are possible.) Somerville and Davies have discussed triangular dissections when no vertex of any triangle falls on the interior of an edge of any other.

H. L. Davies, Packings of spherical triangles and tetrahedra, in [Fen], 42–51; *MR* **36** #3235.

R. D. Mauldin, Problem 60, [Mau].

D. M. Y. Sommerville, Division of space by congruent triangles and tetrahedra, *Proc. Royal Soc. Edinburgh* **43** (1923) 85–116.

M. J. Wenninger, *Spherical Models*, Cambridge University Press, Cambridge, 1979.

C9. The simplexity of the d-cube.

Define the **simplexity**, $S(d)$, of the d-dimensional cube to be the smallest number of (d-dimensional) simplexes into which it can be dissected. Then $S(2) = 2$, and the well-known inscription of a regular tetrahedron in the 3-cube shows $S(3) \leq 5$; in fact equality holds. Several people have shown that $S(4) = 16$, but for $d \geq 5$ the best known results involve rapidly widening bounds: $60 \leq S(5) \leq 67$, and $223 \leq S(6) \leq 344$. Broadie & Cottle have shown that slicing off alternate vertices of the 5-cube leads to 67 or 68 simplexes.

M. N. Broadie & R. W. Cottle, A note on triangulating the 5-cube, *Discrete Math.* **52** (1984) 39–49; *MR* **86c**:52011.

R. W. Cottle, Minimal triangulation of the 4-cube, *Discrete Math.* **40** (1982) 25–29; *MR* **84d**:05065a.

C. W. Lee, Triangulating the *d*-cube, in [GLMP], 205–211; *MR* **87d**:52011.

P. S. Mara, Triangulations for the cube, *J. Combin. Theory Ser. A* **20** (1976) 170–177; *MR* **53** #10624.

J. F. Sallee, A note on minimal triangulations of an *n*-cube, *Discrete Appl. Math.* **4** (1982) 211–215; *MR* **84g**:52019.

J. F. Sallee, The middle-cut triangulations of the *n*-cube, *SIAM J. Algebraic Discrete Methods* **5** (1984) 407–419; *MR* **86c**:05054.

J. F. Sallee, What is the simplicity of the *d*-cube? *Amer. Math. Monthly* **91** (1984) 628–629.

J. F. Sallee, A triangulation of the *n*-cube, *Discrete Math.* **40** (1982) 81–86; *MR* **84d**:05065b.

C10. **Tiling the plane with squares.** One can tile the plane with distinct integer sided squares by starting with any perfect squared square (e.g., one of those in Figure C3) and spiralling outwards with squares of increasing size (see Figure C9). Is it possible to tile the plane with integer squares using exactly one of each size? Figure C10 shows a start, involving the first 28 squares, except for 27. This, or a similar diagram, is due to Golomb. There are various schemes that can tile over three quadrants of the plane using each tile once (see Grünbaum or Gardner) but fitting squares into the remaining region proves to be hard.

In the same vein, is it possible to tile the plane using one square of each rational side?

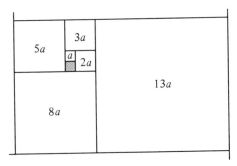

Figure C9. Tiling the plane by integer squares of different sizes. The shaded square is a squared square of side a, such as in Figure C3(a) or (b), and larger squares spiral outwards.

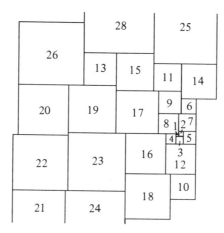

Figure C10.   A compact arrangement of integer squares of sides 1 to 26 and 28. Is it possible to tile the plane using one square of each side?

In three dimensions, Dawson has shown that in every tiling of space with cubes of integer edge lengths, there must be two neighboring cubes of equal size. In particular, it is not possible to tile 3-space with distinct integer-sided cubes.

R. J. Mac G. Dawson, On filling space with different integer cubes, *J. Combin. Theory Ser. A* **36** (1984) 221–229; *M R* **86b**:52008.

D. E. Daykin, Space filling with unequal cubes, *Bull. Amer. Math. Soc.* **70** (1964) 340–341.

M. Gardner, Some packing problems that cannot be solved by sitting on the suitcase, *Scientific American* **241** (Oct. 1979) no. 4, 22–26.

S. Golomb, *J. Recreational Math.* **8** (1975) 138–139.

B. Grünbaum, [Gru], Section 2.4.

C11.   Tiling the plane with triangles. Is there a tiling of the plane by equilateral triangles all of different sizes? The procedure for tiling the plane by different integer sided squares, (see Section C10) does not work here, since no equilateral triangle can be partitioned into unequal equilateral triangles (see Tutte). Scherer shows that any such tiling cannot contain a triangle of minimal side-length.

Is there a tiling of the plane with integer-sided triangles, using just one copy of each such triangle?

Conway realized that the plane can be tiled using just one triangle from each congruence class of rational sided triangles. A constructive demonstration was given by Eggleton, but in general the vertices of some triangles were interior points of sides of others. Is there a strict tiling of the plane with just one triangle from each congruence class of rational triangles, where by "strict" is meant that any common boundary point of two tiles is either a vertex of both tiles or of neither?

Pomerance gave a locally finite tiling (i.e., one in which any disk intersects only finitely many tiles) using the Nagell–Lutz theorem on elliptic curves. He obtains a more elementary solution by sacrificing local finiteness.

Pomerance and Eggleton prove that there is a strict tiling of 3-space, not locally finite, with one rational-edged tetrahedron from each isometry class. Can this be done in a locally finite fashion? For $d > 3$, can you tile d-space with rational edged simplexes, one from each isometry class?

E. Buchman, Tiling with equilateral triangles, *Amer. Math. Monthly* **88** (1981) 748–753.

J. H. Conway, Problem 5328, *Amer. Math. Monthly* **72** (1965) 915, Solution by D. C. Kay, *ibid.* **73** (1966) 903–904.

R. B. Eggleton, Tiling the plane with triangles, *Discrete Math.* **7** (1974) 53–65; *MR* **48** #10977.

R. B. Eggleton, Where do all the triangles go? *Amer. Math. Monthly* **82** (1975) 499–501; *MR* **51** #1584.

N. J. Fine, On rational triangles, *Amer. Math. Monthly* **83** (1976) 517–521; *MR* **54** #2585.

C. Pomerance, On a tiling problem of R. B. Eggleton, *Discrete Math.* **18** (1977) 63–70; *MR* **57** #17515.

K. Scherer, The impossibility of a tesselation of the plane into equilateral triangles whose sidelengths are mutually different, one of them being minimal, *Elem. Math.* **38** (1983) 1–4; *MR* **84f**:52013.

W. T. Tutte, The dissection of equilateral triangles into equilaterial triangles, *Proc. Cambridge Philos. Soc.* **44** (1948) 463–482; *MR* **10**, 319.

W. T. Tutte, Dissections into equilateral triangles, in [Kla], 127–139.

C12. **Rotational symmetries of tiles.** The search for a tiling of the plane in which each tile has 5-fold rotational symmetry goes back at least to Kepler in the early 17th century. Recently, a surprisingly simple tiling was found satisfying these conditions. Enlarge Figure C11(a) to three times its linear dimensions, and replace each of the five pentagons by the original figure.

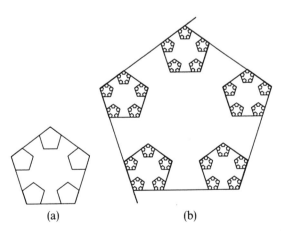

(a)   (b)

Figure C11.   Tiling the plane using tiles with 5-fold symmetry. Each of the small pentagons in (a) is replaced by a scaled copy of the entire figure, and this process is repeated to give a tiling looking locally like (b).

Repeating this process indefinitely leads to the "self-similar" tiling indicated in Figure C11(b). In this example, however, the tiles are not bounded in diameter. Therefore, Danzer, Grünbaum & Shephard ask whether there is a tiling of the plane by tiles which are simply connected, have 5-fold symmetry, and all of which have a diameter of at most 1.

In the same way, are there tilings of the plane with all the tiles having n-fold symmetry for $n = 7, 8, \ldots$? There are obvious examples of tilings with all tiles having 3-, 4-, or 6-fold symmetry.

Grünbaum also asks a question of a rather different nature: if T is a tile with 6-fold and reflective symmetries, such that congruent copies of T tile the plane, is T necessarily a finite union of equal regular hexagons?

H. S. M. Coxeter, *Introduction to Geometry*, Wiley, New York, 1961, Section 4.5; *MR* 23 #A1251.
L. Danzer, B. Grünbaum & G. C. Shephard, Can all tiles of a tiling have five-fold symmetry? *Amer. Math. Monthly* **89** (1982) 568–585.
L. Fejes Tóth, [Fej'].

C13. Tilings with a constant number of neighbors.
L. Fejes Tóth defines $K(n)$ as the least number of convex polygons whose congruent replicas can be put together to form an n-neighbor tiling (i.e., each tile has a boundary point in common with exactly n other tiles). An example with $n = 11$ is shown in Figure C12. His son has proved that for each $n > 5$, $K(n)$ exists and $K(n) \le (n + 1)/2$. It is known that $K(n) = 1$ for $n = 6, 7, 8, 9, 10, 12, 14, 16,$ and 21, and it is conjectured that these are the only values of n for which $K(n) = 1$. Also, $K(n) \le 2$ for $n = 11$ and 13. Find further information about $K(n)$. Does it tend to infinity with n? It is not even certain that $K(n) > 1$ for all sufficiently large n.

The values of n for which $K(n) = 1$ have been determined in the special case where the tiles meeting at each point all have equal angles at the point (see L. Fejes Tóth).

Does $K(n) = 1$ for any further values of n if tilings by non-convex polygons are allowed? It seems unlikely that $K(n) \to \infty$ in this case.

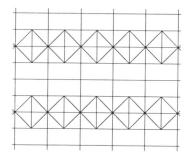

Figure C12.   A tiling in which each tile has exactly 11 neighbors.

K. Bezdek, Mosaics with each face having the same number of neighbors, *Math. Lapok.* **28** (1977/80) 297–303; *M R* **82i**:52023.

K. Böröczky, Über die Newtonsche Zahl regulärer Vielecke, *Period. Math. Hungar.* **1** (1971) 113–119; *M R* **44** #4644.

G. Fejes Tóth, Über Parkettierungen konstanter Nachbarnzahl, *Studia Sci. Math. Hungar.* **6** (1971) 133–135; *M R* **49** #11390.

L. Fejes Tóth, Scheibenpackungen konstanter Nachbarnzahl, *Acta Math. Acad. Sci. Hungar.* **20** (1969) 375–381; *M R* **41** #2536.

L. Fejes Tóth, Remarks on a theorem of R. M. Robinson, *Studia Sci. Math. Hungar.* **4** (1969) 441–445; *M R* **40** #7951.

L. Fejes Tóth, Five-neighbour packing of convex discs, *Period. Math. Hungar.* **4** (1973) 221–229; *M R* **49** #9745,

L. Fejes Tóth, Tesselation of the plane with convex polygons having a constant number of neighbors, *Amer. Math. Monthly* **82** (1975) 273–276; *M R* **51** #1128.

P. Gács, Packing convex sets in the plane with a great number of neighbours, *Acta Math. Acad. Sci. Hungar.* **23** (1972) 383–388; *M R* **48** #7124.

J. Lindhart, Über einige Vermutungen von L. Fejes Tóth, *Acta Math. Acad. Sci. Hungar.* **24** (1973) 199–201; *M R* **47** #2498.

J. Lindhart, Endliche n-Nachbarn packungen in der Ebene und auf der Kugel, *Period. Math. Hungar.* **5** (1974) 301–306; *M R* **51** #11289.

E. Makai, Jr., Five-neighbour packing of convex discs, *Period. Math. Hungar.* **5** (1974).

R. Penrose, The role of aesthetics in pure and applied mathematical research, *Bull. Inst. Math. Appl.* **10** (1974) 266–271.

R. M. Robinson, Finite sets of points on a sphere with each nearest to five others, *Math. Ann.* **179** (1969) 296–318; *M R* **39** #2068.

G. Wegner, Bewegungstabile Packungen konstanter Nachbarnzahl, *Studia Sci. Math. Hungar.* **6** (1971) 431–438; *M R* **45** #4284.

C14. **Which polygons tile the plane?** An old problem is to determine the convex polygons T such that the plane may be tiled with (directly or reflectively) congruent copies of the single tile T. It is straightforward to see that any triangle or quadrilateral may be used. Moreover, a basic theorem states that such tiling polygons can have at most six sides. Thus the problem is to determine which pentagons and hexagons can tile the plane. At present, 14 essentially different classes of pentagon, and 3 classes of hexagon are known to tile the plane (see Grünbaum & Shephard, Chapter 9, and Stein). It has been shown that no further hexagons can be used. Can tilings be found using any other convex pentagons?

A complete analysis of isohedral tilings by congruent copies of a single convex polygon has been given. A tiling is **isohedral** if, given any two of the tiles T_1 and T_2, there is a congruence transformation of the entire tiling onto itself with T_1 mapped onto T_2. There are 107 essentially different isohedral tilings by convex polygons—14 using triangles, 56 using quadrilaterals, 24 using pentagons, and 13 using hexagons (see Grünbaum & Shephard, Section 9.1). Thus, when looking for further polygons, tilings with rather less symmetry will be needed.

The problem of determining which non-convex polygons tile the plane is much harder. Heesch & Kienzle list convex and non-convex hexagons that tile isohedrally—are there any non-convex hexagons that permit a non-isohedral tiling?

Is there an algorithm that will determine within a finite number of steps whether or not a given tile many be used to tile the plane? This seems unlikely—it has been shown that there is no algorithm that will establish whether congruent copies of certain finite collections of tiles can tile the plane (see Berger and Robinson).

R. Berger, The undecidability of the domino problem, *Mem. Amer. Math. Soc.* **66** (1966); *MR* **36** #49.
B. Bollobás, Filling the plane with congruent convex hexagons without overlapping, *Ann. Univ. Sci. Budapest. Sect. Math.* **6** (1983) 117–123.
J. A. Dunn, Tessellations with pentagons, *Math. Gaz.* **55** (1971) 366–369.
B. Grünbaum & G. C. Shephard, [Gru].
B. Grünbaum & G. C. Shephard, Isohedral tilings of the plane by polygons, *Commun. Math. Helv.* **53** (1978) 542–571; *MR* **80d**:52016.
B. Grünbaum & G. C. Shephard, Some problems on plane tilings, in [Kla], 167–196.
H. Heesch & O. Kienzle, *Flächenschluss. System der Formen lückenlos aneinander schliessender Flachteile*, Springer, Berlin, 1963.
M. D. Hirschhorn & D. C. Hunt, Equilateral convex pentagons which tile the plane, *J. Combin. Theory Ser. A.* **39** (1985) 1–18; *MR* **86g**:52022.
R. B. Kershner, On paving the plane, *Amer. Math. Monthly* **75** (1968) 839–844; *MR* **38** #5116.
R. M. Robinson, Undecidability and nonperiodicity for tilings in the plane, *Invent. Math.* **12** (1971) 177–209; *MR* **45** #6626.
D. Schattschneider, Tiling the plane with congruent pentagons, *Math. Mag.* **51** (1978) 29–44; *MR* **58** #12735.
D. Schattschneider, In praise of amateurs, in [Kla], 140–166.
R. Stein, A new pentagon tiler, *Math. Mag.* **58** (1985) 308.

C15. Isoperimetric problems for tilings.

Let T be a tile of unit area such that the plane may be tiled by congruent copies of it. Steinhaus asks if the perimeter length of T is least when T is a regular hexagon. More generally, if the plane is tiled by bounded tiles, not necessarily congruent, but all of a diameter of at least D_0, say, does the regular hexagonal tiling minimize the maximum of (perimeter length of $T)^2$/(area of T) taken over all tiles T in the tiling?

The 3-dimensional analog of this is likely to be challenging: What is the tile T of unit volume and least surface area that permits a tiling of $\mathbb{R}^3$ by congruent copies of T?

C16. Polyominoes.

An n-omino, or **polyomino** if we do not wish to mention n explicitly, is the union of n squares from a square grid joined edge to edge to form a simply connected block (i.e., a block without holes). The term polyomino was coined in 1953 by Golomb, whose book provides an excellent and copiously illustrated introduction to the subject. The n-ominoes for $n \leq 5$ are shown in Figure C13 (for larger n see Grünbaum & Shephard).

A first problem is to determine how many n-ominoes there are for each n. For small n:

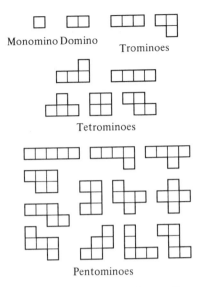

Figure C13.   The polyominoes of up to five squares.

n:	1	2	3	4	5	6	7	8	9	10	11	12	13	14
number:	1	1	2	5	12	35	107	369	1285	4655	17073	63600	238591	901971

Redelmeier continues the table up to $n = 24$; substantial computation is required for the larger values of n. What is a good asymptotic approximation to the number of n-ominoes for large n? Conway claims ca^n where $4 < a \le 6.75$. Klarner & Rivest and Delest consider this for restricted classes of polyominoes [e.g., when each column (or row) consists of a connected block of squares].

There is no end to the list of problems on tiling regions by polyominoes. The **order** of a polyomino is the smallest number of (directly or reflectively) congruent copies that can be arranged to form a rectangle. What is the order of the small polyominoes, and in particular, which of them have finite order (i.e., which tile *some* rectangle)? All but one of the tetrominoes have finite order, but only five of the pentominoes and ten of the hexominoes do so. It was only established recently, after extensive computation by Dahlke, that the Y-hexomino (5 squares in a row with the 6th attached one square from the end) has order 92, the highest finite order of any hexomino by a factor of more than 5.

Golomb (1989) considers the question of which orders can actually occur. He shows that there are infinitely many examples of polyominoes of order $4n$ for each positive integer n. It is easy to see that order 2 can occur, and Klarner has given examples with orders 10 and 18. What other orders are possible; in particular is there any polyomino of odd order greater than one? Eleven copies of a certain hexomino (three quarters of a 2×4 rectangle) may be arranged to form a rectangle (see Figure C7) though, of course, this is not the minimal number. Klarner asks if there is any (non-rectangular) polyomino 3, 5, 7, or 9 copies of which tile a rectangle. Berger has shown that the problem of whether a given polyomino can tile a rectangle is computationally undecidable. In particular, there is no computable function $f(n)$ that bounds the area of the smallest rectangle that an n-omino can tile.

The question of which polyominoes tile the plane is also of great interest. Clearly, if a polyomino tiles a rectangle, then the tiling may be extended to the plane. However, other tilings may be possible. It is possible to tile the plane using congruent copies of any n-omino with $n \le 6$. However, there are 3 heptominoes and 20 octominoes that do not tile the plane. For $n \ge 9$ there are many n-ominoes for which it is unknown whether tilings are possible. There are various necessary or sufficient conditions for a polyomino to tile the plane (see Golomb or Gardner), but there are many polyominoes to which these fail to apply. Barnes has given a powerful algebraic method of determining whether a given set of tiles tile the plane in a generalized sense.

Of course, one can ask whether given sets of polyominoes can be used to tile various regions. In particular, for each n, which regions can be tiled using a complete set of n-ominoes? For example, the set of pentominoes can be arranged into 6×10, 5×12, or 4×15 rectangles, the heptominoes can tile a 7×107 rectangle, but there is a simple argument to show that no rectangle can be tiled using the set of 35 hexominoes (see Golomb's book for a discussion of such problems).

There are many variations on polyominoes. Similar questions can be asked for the **polyiamonds**, formed by triangles taken from the regular triangular grid, or for the **polyhexes**, obtained using the regular hexagonal grid. Alternatively, we can allow the shapes to meet corner-to-corner as well as edge-to-edge, or drop the requirement of simply-connectedness. There are also generalizations to higher dimensions: model-makers may care to experiment with the **polyominoids** obtained by glueing a collection of unit cubes face-to-face in some arrangement.

F. W. Barnes, Algebraic theory of brick packing I, II, *Discrete Math.* **42** (1982) 7–26, 129–144; *MR* **84e**:05044a,b.

E. Bender, Convex n-ominoes, *Discrete Math.* **8** (1974) 219–226; *MR* **49** #67.

R. Berger, The undecidability of the domino problem, *Mem. Amer. Math. Soc.* **66** (1966); *MR* **36** #49.

K. A. Dahlke, The Y-hexomino has order 92, *J. Combin. Theory Ser. A* **51** (1989) 125–126.

K. A. Dahlke, A heptomino of order 76, *J. Combin. Theory Ser. A* **51** (1989) 127–128.

T. R. Dawson & W. E. Lester, A notation for dissection problems, *Fairy Chess Review* **3** #5 (Apr. 1937) 46–47.

M.-P. Delest, Generating functions for column-convex polyominoes, *J. Combin. Theory Ser. A* **48** (1988) 12–31.

M.-P. Delest & G. Viennot, Algebraic languages and polyominoes enumeration, *Theoret. Comput. Sci.* **34** (1984) 169–206; *MR* **86e**:68062.

M. Gardner, More about tiling the plane: the possibility of polyominoes, polyiamonds and polyhexes, *Sci. American* **233**, no. 2 (Aug. 1975) 112–115.

M. Gardner, *New Mathematical Diversions from Scientific American*, Simon and Schuster, New York, 1966.

S. W. Golomb, *Polyominoes*, Allen and Unwin, London, 1966.

S. W. Golomb, Tiling with polyominoes, *J. Combin. Theory* **1** (1966) 280–296; *M R* **33** #6498.

S. W. Golomb, Tiling with sets of polyominoes, *J. Combin. Theory* **9** (1970) 60–71; *M R* **41** #93.

S. W. Golomb, Polyominoes which tile rectangles, *J. Combin. Theory Ser. A* **51** (1989) 117–124.

B. Grünbaum & G. C. Shephard, [GS], Section 9.4.

J. Haselgrove, Packing a square with Y-pentominoes, *J. Recreational Math.* **7** (1974) 229.

M. S. Klamkin & A. Liu, Polyominoes on the infinite checkerboard, *J. Combin. Theory Ser. A* **28** (1980) 7–16; *M R* **81h**:05047.

D. A. Klarner, Covering a rectangle with L-tetrominoes, *Amer. Math. Monthly* **70** (1963) 760.

D. A. Klarner, Some results concerning polyominoes, *Fibonacci Quart.* **3** (1965) 9–20; *M R* **32** #4028.

D. A. Klarner, Packing a rectangle with congruent *n*-ominoes, *J. Combin. Theory* **7** (1969) 107–115; *M R* **40** #1894.

D. A. Klarner, My life among the polyominoes, in [Kla], 243–262.

D. A. Klarner & F. Göbel, Packing boxes with congruent figures, *Indagationes Math.* **31** (1969) 465–472; *M R* **40** #6362.

D. A. Klarner & R. L. Rivest, A procedure for improving the upper bound for the number of *n*-ominoes, *Canad. J. Math.* **25** (1973) 585–602; *M R* **48** #1943.

D. A. Klarner & R. L. Rivest, Asymptotic bounds for the number of convex *n*-ominoes, *Discrete Math.* **8** (1974) 31–40; *M R* **49** #91.

W. F. Lunnon, Counting polyominoes, in *Computers in Number Theory*, Academic Press, London, 1971, 347–372.

R. C. Read, Contributions to the cell growth problem, *Canad. J. Math.* **14** (1962) 1–20; *M R* **24** #A1219.

D. H. Redelmeier, Counting polyominoes; yet another attack, *Discrete Math.* **36** (1981) 191–203; *M R* **84g**:05049.

D. Walkup, Covering a rectangle with T-tetrominoes, *Amer. Math. Monthly* **72** (1965) 986–988; *M R* **32** #1582.

C17. **Reptiles.** An *n*-**reptile** is a two dimensional region that can be tiled by *n* congruent tiles, each similar to the whole region. We look for reptiles with a reasonably simple topological structure—bounded tiles that are the closure of their interior, say. Any such reptile may be used to tile the plane by repeated subdivision and rescaling (see Figure C14). (It is this replication process that gives reptiles their name.)

To begin with, we seek reptiles that are convex polygons. The only such 2-reptiles are parallelograms with sides in the ratio $1 : \sqrt{2}$ and the right-angled isosceles triangle (see Figure C15). More generally, a parallelogram with sides of ratio $1 : \sqrt{n}$ is an *n*-reptile. The body of theory on monohedral tilings of the plane (see Grünbaum & Shephard) should make a complete classification of convex polygonal reptiles possible; in particular such a reptile cannot have more than six sides.

Analysis of non-convex polygons that are *n*-reptiles may be rather harder, though a characterization might be possible for small *n*, say, $n = 2, 3,$ and 4. Some examples are shown in Figure C16. In particular, which polyominoes are reptiles (see Section C16)?

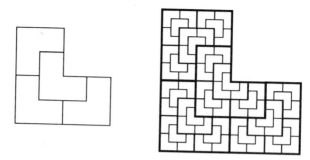

Figure C14.   How a reptile leads to a tiling of the plane.

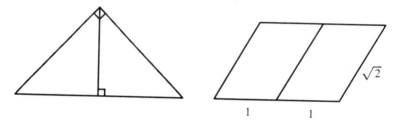

Figure C15.   Convex 2-reptiles.

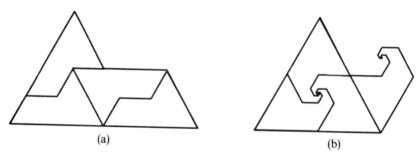

(a)   (b)

Figure C16.   4-reptiles: (a) polygonal and (b) spiral.

There are various curious reptiles that are bounded by a countable number of straight line segments [see Figure C16(b) for an example].

Are there any 2-reptiles other than the convex ones with "reasonably simple" boundaries?

Dekking gives a general method for generating a large class of reptiles called **fractiles**. These are reptiles whose boundaries are fractal curves.

John Conway asked if there are any "reptiles with holes," (i.e., that are not simply-connected). Grünbaum gave the example of Figure C17 which is a 36-reptile, and suggested that the proper question to ask is whether there is a connected open set whose closure is a reptile that is not simply connected. What is the least n for which there is an n-reptile with any sort of hole?

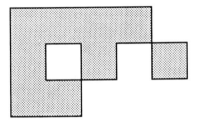

Figure C17. Grünbaum's 36-reptile with a hole (note that two copies fit together to form a 3 × 6 rectangle).

Grünbaum also asks whether there is any 2-reptile that is also a 3-reptile. More generally, for which n and m do there exist n-reptiles that are also m-reptiles?

We might also ask to characterize those sets which may be tiled by n tiles not necessarily the same size, but each *similar* to the original.

Self-similar fractals give examples of reptiles with empty interiors, and these have been studied intensively (see Barnsley or Falconer). The simplest example is the middle-third Cantor set, which is made up of two copies of itself scaled by a factor of $\frac{1}{3}$.

M. Barnsley, *Fractals Everywhere*, Academic Press, New York, 1988.

R. O. Davies, Replicating boots, *Math. Gaz.* **50** (1966) 175.

F. M. Dekking, Replicating superfigures and endomorphisms of free groups, *J. Combin. Theory Ser. A* **32** (1982) 315–320; *MR* **84i**:51019.

F. M. Dekking, Recurrent sets, *Adv. Math.* **44** (1982) 78–104; *MR* **84e**:52023.

J. Doyen & M. Landuyt, Dissections of polygons, *Ann. Discrete Math.* **18** (1983) 315–318.

K. J. Falconer, *Fractal Geometry–Mathematical Foundations and Applications*, Wiley, Chichester, 1990.

J. Giles, Jr., Infinite-level replicating dissections of plane figures, *J. Combin. Theory Ser. A* **26** (1979) 319–327; *MR* **80g**:51013a.

J. Giles, Jr., Construction of replicating superfigures, *J. Combin. Theory Ser. A* **26** (1979) 328–334; *MR* **80g**:51013b.

J. Giles, Jr., Superfigures replicating with polar symmetry, *J. Combin. Theory Ser. A* **26** (1979) 335–337; *MR* **80g**:51013c.

M. Goldberg & B. M. Stewart, A dissection problem for sets of polygons, *Amer. Math. Monthly* **71** (1964) 1077–1095; *MR* **30** #508.

S. W. Golomb, Replicating figures in the plane, *Math. Gaz.* **48** (1964) 403–412.

G. Valette & T. Zamfirescu, Les partages d'un polygone convexe en 4 polygones semblables au premier, *J. Combin. Theory Ser. B* **16** (1974) 1–16; *MR* **48** #9551.

C18. **Aperiodic tilings.** A tiling of the plane is **periodic** if the tiling may be translated onto itself in two nonparallel directions. A set of prototiles (see the Chapter C introduction) is called **aperiodic** if congruent copies of the prototiles admit infinitely many tilings of the plane, none of which are periodic. It must be emphasized that no periodic tilings at all are permitted even using just one of the prototiles. For example, a 2 × 1 rectangle can tile the plane in a nonperiodic manner, but obvious periodic tilings are also possible.

The first set of aperiodic tiles, consisting of 20,426 prototiles, was discovered by Berger in 1966. Berger managed to reduce this number to 104, and others found successively smaller sets of aperiodic tiles, with Robinson constructing an aperiodic set of six tiles in 1971. The most remarkable discovery was by Penrose in 1974, who gave a set of just two aperiodic quadrilaterals [Figure C18(a)]. In fact, to get aperiodicity, a matching condition must be imposed, so that vertices always meet with the same color. However, this matching condition may be imposed geometrically, by modifying the sides slightly as indicated in Figure C18(b). A tiling of a portion of the plane by these "Penrose Pieces" is shown in Figure C19. While it is not too difficult to show that certain prototiles are aperiodic, analysis of the form of possible tilings is complicated. For example, every bounded region of tiles in a tiling by Penrose Pieces is congruent to the tiling in infinitely many regions in any other such tiling. Moreover, any tiling of the plane uses the two tiles in the proportion $1 : (1 + \sqrt{5})/2$, the golden ratio.

The big problem is to find sets of aperiodic prototiles that are essentially different from the ones already known. In particular, is there a single aperiodic prototile (with or without a matching condition), that is, one that admits only aperiodic tilings by congruent copies? There is a set of three convex polygons that are aperiodic with no matching condition on the edges, but are two or even one such convex prototiles possible?

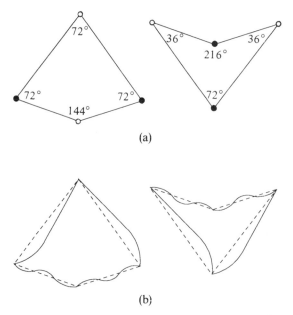

(a)

(b)

Figure C18. (a) The Penrose aperiodic tiles which must be arranged so that like colored vertices meet. (b) By distorting the edges this matching condition may be encoded in the shape of the tiles.

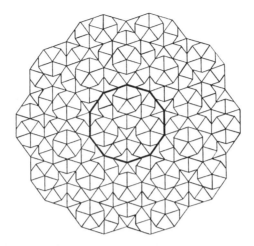

Figure C19.   Part of a tiling of the plane by Penrose tiles. The tiling may be built outwards in a series of concentric 'cartwheels'.

We can also try to find information about the tilings that are possible using aperiodic tiles. What symmetry groups can tilings have? What proportions can the different shaped tiles be used in?

Of course, we may ask similar questions in three dimensions (where a tiling is periodic if it may be translated onto itself in three linearly independent directions) or higher dimensions. In three dimensions Ammann has discovered two aperiodic polyhedra, and Danzer has found a family of four aperiodic tetrahedra. Aperiodic tilings in three dimensions have recently attracted enormous interest following the discovery of their physical importance as quasicrystals. For details of these aspects, see Gratias & Michel or Jaric or Steinhardt & Ostland.

For detailed accounts of aperiodic tilings and many further references see the books by Gardner and Grünbaum & Shephard.

L. Danzer, Three-dimensional analogues of the planar Penrose tilings and quasi-crystals, *Discrete Math.* **76** (1989) 1–7; *MR* **90g**:52015.

M. Gardner, *Penrose Tiles to Trap-door Cyphers*, W. H. Freeman, New York, 1989, Chapters 1 and 2.

D. Gratias & L. Michel (eds.), International Workshop on Aperiodic Crystals, *J. Physique Colloq.* C3 Supp. **7** (1986); *MR* **87h**:82004.

B. Grünbaum & G. C. Shephard, [GS], Chapters 10 and 11.

M. V. Jaric (ed.), *Introduction to the Mathematics of Quasicrystals*, Academic Press, San Diego, 1989.

P. J. Steinhardt & S. Ostland (eds.), *The Physics of Quasicrystals*, World Scientific, Singapore, 1987.

C19. **Decomposing a sphere into circular arcs.** Mikusiński asked whether the surface of a sphere could be expressed as a disjoint union of congruent non-overlapping great circle arcs. Here, congruence implies not only equality of length, but also of topological type, thus there are three cases to consider: (1) open, (2) half open, and (3) closed arcs. Conway & Croft showed the first case to be impossible and the second case (easily) possible. The closed case has resisted all attempts; we guess, rather tentatively, that the answer is negative, although (i) slight weakening of the equal-length condition leads to a possible configuration, and (ii) in the analogous problem of covering the plane with unit segments there *is* a configuration of closed segments; the non-compactness of the plane being used very strongly—essentially "we put all the difficulties at infinity." The open and half-open cases in the plane are settled similarly to those on the sphere.

The following conjecture arises from considering the above: if a sphere (or plane) is covered with non-overlapping arcs, each differentiable, rectifiable and of length at least $\delta > 0$, then at almost all points x, the covering is "smooth" in the sense that given $\varepsilon > 0$ there is a neighborhood N of x containing no endpoint of an arc, and with all the tangents to all the arcs in N having directions differing by less than ε.

In 1939, Knaster recorded in the Scottish Book the problem of whether there was an effective (i.e., not using the axiom of choice) decomposition of the (solid) unit ball into disjoint chords. Apparently this problem is still open, though it seems unlikely that the prize of "a small light beer" is still on offer!

J. H. Conway & H. T. Croft, Covering a sphere with congruent great-circle arcs, *Proc. Cambridge Philos. Soc.* **60** (1964) 787–800; *MR* **29** #6394.
B. Knaster, Problem 182, in [Mau].

C20. Problems in equidecomposability.
Two subsets E and F of $\mathbb{R}^d$ ($d \geq 1$) are (**finitely**) **equidecomposable** under congruence, resp. translation, if they can be expressed as finite disjoint unions

$$E = \bigcup_{i=1}^{k} E_i, \qquad F = \bigcup_{i=1}^{k} F_i,$$

where E_i is congruent to, resp. a translate of, F_i for each i. If the decompositions are into closed topological disks bounded by Jordan curves with merely the interiors of the discs disjoint, we say E and F are **scissors equidecomposable**. (Imagine cutting up E and rearranging the pieces to get F.) If the unions are countable rather than finite, we refer to **countable equidecomposability**.

The basic results go back to Banach & Tarski who proved that any two bounded sets in $\mathbb{R}^d$ ($d \geq 1$) that contain interior points are countably equi-decomposable. The same is true for finite equidecomposability in $\mathbb{R}^d$ ($d \geq 3$); in particular we have the famous Banach–Tarski paradox, that a solid ball may be divided into five pieces which may be rearranged to form two solid balls each congruent to the original one. They also showed that such results hold on the surface of the sphere, but that two measurable sets in $\mathbb{R}^1$ or $\mathbb{R}^2$ can only be finitely equidecomposable if they are of equal Lebesgue measure.

In 1924 Tarski showed that two polygons are finitely equidecomposable, if and only if they have equal area. This led him to pose the most intriguing question of whether a square and circle of equal area are finitely equide-composable. Recently this was solved affirmatively in a remarkable paper by Laczkovich. Moreover, he showed that a square can be decomposed into a finite number of pieces that can be rearranged using *translations only* to form a circle, or, indeed, any set of the same area with piecewise smooth boundary of bounded curvature. Even more recently, and more remarkably, Laczkovich showed that any pair of bounded measurable subsets of $\mathbb{R}^d$ of equal d-dimensional content are translation equidecomposable provided that their boundaries are 'not too large'. In particular, any two plane sets of equal area with rectifiable boundaries are translation equidecomposable, as are a ball and cube of equal volume in $\mathbb{R}^3$. These constructions depend heavily on the axiom of choice—can this be dispensed with, perhaps with the pieces Lebesgue measurable, or even Borel sets? How many pieces are required in the decomposition? Laczkovich estimates that his construction for squaring the circle requires about 10^{50} pieces. It is not hard to see that at least three pieces are required; can closer bounds be obtained?

A famous 3-dimensional result, the Dehn–Sydler theorem, gives neces-sary and sufficient conditions for two polyhedra to be equivalent by decom-position into a finite number of *polyhedra*. Dehn gave necessary conditions in 1901, which in particular answered Hilbert's third problem by showing that there exist non-equidecomposable pyramids on the same base and of the same height. Much more recently Sydler proved that Dehn's conditions were sufficient. In a series of papers, Hadwiger has generalized these condi-tions to higher dimensional polyhedra, but the necessary conditions have only been shown sufficient in $\mathbb{R}^4$ (see Jessen); precise conditions in five or more dimensions have not yet been proved. Hadwiger also considered this problem with translation replacing congruence.

Dubins, Hirsch & Karush have characterized the pairs of plane con-vex sets that are scissors equidecomposable. In addition to the sets having equal area, their boundaries must also be equidecomposable in a certain 1-dimensional sense. Sallee showed that plane sets that are scissors equide-composable are convex equidecomposable (i.e., the parts can be chosen to be convex) but recently Gardner gave an example of a pair of equidecomposable convex sets in the plane that were not equivalent under any decomposition into convex parts (and thus not scissors equidecomposable).

Another interesting problem is due to de Groot (see Dekker). Can the Banach–Tarski transformation of a unit ball into two unit balls be affected "physically" (i.e., can the pieces be moved from one configuration to the other without interpenetration)?

There is much literature on this fascinating subject. For further reading and problems see the classic account by Sierpiński, or the more recent books by Boltyanskiĭ, by Sah, and by Wagon.

S. Banach & A. Tarski, Sur la décomposition des ensembles de points en parties respectivement congruents, *Fund. Math.* **6** (1924) 244–277.

V. G. Boltyanskiĭ, *Equivalent and Equidecomposable Figures*, Heath, Boston, 1963.

V. G. Boltyanskiĭ, *Hilbert's Third Problem*, Winston, Washington, 1978.

M. Dehn, Über den Rauminhalt, *Math. Ann.* **55** (1901).

T. J. Dekker, Paradoxical decompositions of sets and spaces, thesis, University of Amsterdam, 1958; *MR* **22** #12430.

T. J. Dekker & J. de Groot, Decompositions of a sphere, *Fund. Math.* **43** (1956) 185–194; *MR* **19**, 1068.

L. Dubins, M. W. Hirsch & J. Karush, Scissor congruence, *Israel J. Math.* **1** (1963) 239–247; *MR* **29** #2706.

R. J. Gardner, A problem of Sallee on equidecomposable convex bodies, *Proc. Amer. Math. Soc.* **94** (1985) 329–332; *MR* **86f**:52005.

R. J. Gardner, Convex bodies, equidecomposable by locally discrete groups of isometries, *Mathematika* **32** (1985) 1–9; *MR* **87c**:52007.

H. Hadwiger, Zum Problem der Zerlegungsgleichheit k-dimensionaler Polyeder, *Math. Ann.* **127** (1954) 170–174; *MR* **15**, 691.

H. Hadwiger, Polytypes and translative equidecomposability, *Amer. Math. Monthly* **79** (1972) 275–276.

H. Hadwiger & P. Glur, Zerlegungsgleichheit ebener Polygone, *Elem. Math.* **6** (1951) 97–106; *MR* **13**, 576.

B. Jessen, The algebra of polyhedra and the Dehn–Sydler theorem, *Math. Scand.* **22** (1968) 241–256; *MR* **40** #4860.

B. Jessen, Zur Algebra der Polytope, *Nachr. Akad. Wiss. Göttingen Math. Phys. Kl.* **II** (1972) 47–53; *MR* **50** #5636.

M. Laczkovich, Equidecomposability and discrepancy: a solution to Tarski's circle squaring problem, *J. Reine Angew. Math.* **404** (1990) 77–117.

M. Laczkovich, Decomposition of sets with small boundary, *J. London Math. Soc.*, to appear.

P. McMullen & R. Schneider, Valuations on convex bodies, in [GW], 170–247; *MR* **85e**:52001.

V. G. Pokrovskiĭ, On the triangulations and the equidecomposability of n-dimensional parallelepipeds, *Soobshch. Akad. Nauk Gruzin. SSR* **123** (1986) 485–487; *MR* **88f**:52024.

R. Schneider, Equidecomposable polyhedra, in [BF], 481–501; *MR* **89c**:52007.

C. H. Sah, *Hilbert's Third Problem: Scissors Congruence*, Pitman, San Francisco, 1979; *MR* **81g**:51001.

G. T. Sallee, Are equidecomposable plane convex sets convex equidecomposable? *Amer. Math. Monthly* **76** (1960) 926–927.

J.-P. Sydler, Conditions nécessaires et suffisantes pour l'équivalence des polyèdres de l'espace euclidien à trois dimensions, *Comment. Math. Helv.* **40** (1965) 43–80; *MR* **33** #632.

W. Sierpiński, On the congruence of sets and their equivalence by finite decomposition, in *Congruence of Sets and Other Monographs*, Chelsea, New York, 1967; *MR* **15**, 691.

K. Stromberg, The Banach-Tarski paradox, *Amer. Math. Monthly* **86** (1979) 151–161.

A. Tarski, Orównoważnóśei wielokatów Przeglad, *Mate.-Fiz.* **1–2** (1924) 54.

S. Wagon, *The Banach-Tarski Paradox*, Cambridge University Press, Cambridge, 1985; *MR* **87e**:04007.

D. Packing and Covering

When is it possible to pack the sets X_1, X_2, ... into a given "container" X? This is the typical form of a packing problem; we seek conditions on the sets such that disjoint congruent copies (or perhaps translates) of the X_i may be **packed** inside X. Usually we permit boundary contact between the sets. Clearly, for any packing to be possible, the sum of the areas (or volumes) of the X_i cannot exceed that of X, but generally the geometry of the sets results in considerable wasted space in any packing.

In the same way, we might ask when a collection of sets X_1, X_2, ... can be placed so that they **cover** a given set X, with overlapping allowed.

Packing and covering usually go side-by-side. There are dual covering problems to many packing problems, for example, the question of how many disks of radius r can cover a unit square is an obvious dual to the Tammes problem of how many disks of radius r can be packed into a unit square without overlap. Often, similar techniques may be used in both packing and covering problems.

Problems involving packing or covering with disks or spheres are often expressed in terms of arranging points so they are not too close or too far from each other. For example, packing unit disks in the plane is equivalent to arranging a set of points (the centers) so that the distance between any pair is at least 1. It will sometimes be convenient to give such alternative formulations of problems.

This chapter and the previous one are closely related, since a tiling may be regarded as a packing that completely fills the space available or, alternatively, as a covering without overlaps.

J. H. Conway & N. J. A. Sloane, [CS].
G. Fejes Tóth, New results in the theory of packing and covering, [GW], 318–359; *MR* **85i**:52007.

L. Fejes Tóth, [Fej].
F. Göbel, Geometrical packing and covering problems, in *Packing and Covering in Combinatorics*, A. Schrijver (ed.), *Math. Centrum Tracts* **106** (1979) 179–199.
C. A. Rogers, [Rog].

D1. Packing circles or spreading points in a square.

What is the maximum diameter of n equal circles that can be packed into a unit square? How should n points be arranged in a unit square so that the minimum distance between them is greatest? These two problems are equivalent, since if a collection of points in a unit square are a distance of at least d from each other, the points can serve as the centers of a collection of circles of diameter d that will pack into a square of side $1 + d$.

We consider the "spreading points" version of the problem and write d_n for the greatest possible minimum distance between n points in a unit square.

Exact results are known for $n \leq 9$ and $n = 14, 16, 25$, and 36. For $2 \leq n \leq 5$ these are easy to obtain. Graham has established the result for $n = 6$; a simple proof would be of interest. For $n = 7, 8$, and 9 the results are due to Schaer (& Meir). Wengerodt (& Kirchner) solved the problem for $n = 14, 16, 25$, and 36. Goldberg has given conjectural configurations for $n \leq 27$, Schaer, Schlüter, Valette, and Grünbaum have given improved configurations for $n = 10$, and Schlüter for $n = 13$. Current conjectural bounds are listed in Table D1, with conjectural extremals shown in Figure D1.

Up to which square number is the square lattice packing the best? Certainly for up to 36. Wengerodt has found a packing of 64 points that is denser than the square lattice packing, but he conjectures that for 49 the square lattice packing is best.

Table D1. d_n is the greatest minimal separation possible between n points in the unit square.

n	d_n		n	d_n	
2	$\sqrt{2}$	$= 1.414...$*	15	$4/(8 + \sqrt{2} + \sqrt{6}) = 0.337...$	
3	$\sqrt{6} - \sqrt{2}$	$= 1.035...$*	16	$1/3$	$= 0.333...$*
4	1	$= 1$ *	17		$= 0.306...$
5	$\sqrt{2}/2$	$= 0.707...$*	18	$\sqrt{13}/12$	$= 0.300...$
6	$\sqrt{13}/6$	$= 0.601...$*	19		$= 0.290...$
7	$2(2 - \sqrt{3})$	$= 0.536...$*	20	$(6 - \sqrt{2})/16$	$= 0.287...$
8	$(\sqrt{6} - \sqrt{2})/2 = 0.518...$*		21		$= 0.272...$
9	$1/2$	$= 0.5$ *	22	$2 - \sqrt{3}$	$= 0.268...$
10		$= 0.421...$	23	$(\sqrt{6} - \sqrt{2})/4$	$= 0.259...$
11		$= 0.398...$	24	$2/(4 + \sqrt{2} + \sqrt{6}) = 0.254...$	
12	$\sqrt{34}/15$	$= 0.389...$	25	$1/4$	$= 0.25...$*
13	$(\sqrt{3} - 1)/2$	$= 0.366...$	26		$= 0.239...$
14	$(\sqrt{6} - \sqrt{2})/3 = 0.345...$*		27	$\sqrt{89}/40$	$= 0.236...$

* exact values

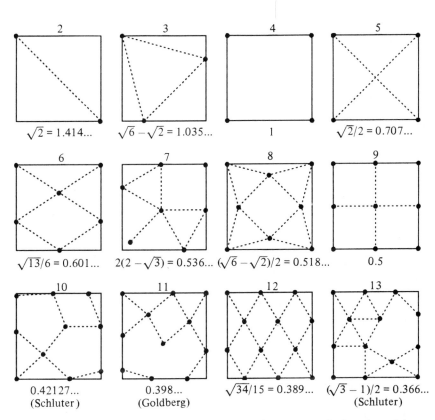

Figure D1.   Arrangements of n points in the unit square such that the minimum distance apart in as large as possible. The configurations shown are optimal for $n \leq 9$.

Are there any values of n such that $d_n = d_{n+1}$? It is easy to see that $d_n \sim 2^{1/2}3^{-1/4}n^{-1/2}$ for large n, but what are good bounds for the error term?

The problem can be asked for packing an equilateral triangle. Oler has shown that if n is triangular (i.e., of the form $m(m+1)/2$) the obvious configuration is the extremal one. Can one ever do better if n is 1 less than a triangular number?

Schaer, Golser, Goldberg, and Bezdek have also packed n spheres into a cube and certain other polyhedra. Even for the cube, exact results are known only for $n \leq 10$.

P. Ament, Kugelpackungen in Konvexen Polytogen, Diplomarbeit, Universität Stuttgart, 1988.

K. Bezdek, Densest packing of a small number of congruent spheres in polyhedra, *Ann. Univ. Sci. Budapest. Eötvös Sect. Math.* **30** (1987) 177–194; *MR* **89a**:52035.

M. Goldberg, The packing of equal circles in a square, *Math. Mag.* **43** (1970) 24–30.

M. Goldberg, On the densest packing of equal spheres in a cube, *Math. Mag.* **44** (1971) 199–208; *MR* **45** #7614.

G. Golser, Dichteste Kugelpackungen im Oktaeder, *Studia Sci. Math. Hungar.* **12** (1977) 337–343; *MR* **82e**:52017.

B. Grünbaum, An improved packing of ten equal circles in a square, to appear.

K. Kirchner & G. Wengerodt, Die dichteste Packung von 36 Kreisen in einem quadrat, *Beiträge Algebra Geom.* **25** (1987) 147–159; *MR* **88g**:52010.

J. H. Lindsey II, Sphere packing in R^3, *Mathematika* **33** (1986) 137–147; *MR* **87m**:52026.

W. Moser & J. Pach, [MP], Problem 6.

N. Oler, A finite packing problem, *Canad. Math. Bull.* **4** (1961) 153–155; *MR* **24** #2901.

J. Schaer, The densest packing of 9 circles in a square, *Canad. Math. Bull.* **8** (1965) 273–277; *MR* **31** #6164.

J. Schaer, The densest packing of five spheres in a cube, *Canad. Math. Bull.* **9** (1966) 271–274; *MR* **34** #684.

J. Schaer, The densest packing of six spheres in a cube, *Canad. Math. Bull.* **9** (1966) 275–280; *MR* **34** #1929.

J. Schaer, On the densest packing of spheres in a cube, *Canad. Math. Bull.* **9** (1966) 265–270; *MR* **34** #683.

J. Schaer, On the packing of ten equal circles in a square, *Math. Mag.* **44** (1971) 139–140.

J. Schaer, The densest packing of ten congruent spheres in a cube, to appear.

J. Schaer & A. Meir, On a geometric extremum problem, *Canad. Math. Bull.* **8** (1965) 21–27; *MR* **30** #5215.

K. Schlüter, Kreispackung in Quadraten, *Elem. Math.* **34** (1979) 12–14; *MR* **80c**:52014.

M. Schmitz & K. Kirchner, Eine Verteilung von 13 Punkten auf einem Quadrat, *Wissensch. Zeit. Pädagog Hoch. Erfurt-Muhlhausen* **18** (1982) 113–115.

G. Valette, A better packing of ten equal circles in a square, *Discrete Math.* **76** (1989) 57–59; *MR* **90g**:52019.

G. Wengerodt, Die dichteste Packung von 16 Kreisen in einem Quadrat, *Beiträge Algebra Geom.* **16** (1983) 173–190; *MR* **85j**:52024.

G. Wengerodt, Die dichteste Packung von 25 Kreisen in einem Quadrat, *Ann. Univ. Sci. Budapest Eötvös Sect. Math.* **30** (1987) 3–15; *MR* **89a**:52041.

G. Wengerodt, Die dichteste Packung von 14 Kreisen in einem Quadrat, *Beiträge Algebra Geom.* **25** (1987) 25–46; *MR* **88g**:52014.

D2. Spreading points in a circle.

D2. **Spreading points in a circle.** The analog of the previous problem for the circle may be stated in various essentially equivalent ways.

What is the maximum radius of a disk, n copies of which can be packed into a circle of radius 1?

What is the radius of the smallest circle into which n unit disks may be packed?

What is the radius of the smallest circle containing n points, no pair being a distance of less than 1 apart?

How large can the least distance between a pair chosen from n points in the unit circle be?

Adopting the last form of the problem, we have that for $2 \le n \le 6$ the least distance is $2 \sin \pi/n$ and for $7 \le n \le 9$ is $2 \sin \pi/(n - 1)$, with the obvious configurations. This is easy to show for $n \le 7$, and was proved by Pirl for $n = 8$ and 9, who also solved the case of $n = 10$, and conjectured the values for $n \le 19$.

M. Goldberg, Packing of 14, 16, 17 and 20 circles in a circle, *Math. Mag.* **44** (1971) 134–139.

S. Kravitz, Packing cylinders into cylindrical containers, *Math. Mag.* **40** (1967) 65–71.

U. Pirl, Der Mindestabstand von n in der Einheitskreisschiebe gelegten Punkten, *Math. Nachr.* **40** (1969) 111–124; *MR* **40** #6379.

D3. Covering a circle with equal disks.

The problem of completely covering a circular region by placing over it, one at a time, five smaller equal circular disks was familiar to frequenters of English fairs a century ago. It can be done if the radius of the smaller disks exceed 0.609383... of that of the circular region. For a discussion of Neville's solution, see Rouse Ball (though the number is given incorrectly in some editions). What is the minimum radius for coverings by other numbers of equal disks? The cases of three, four and seven disks are easy, and Bezdek gives solutions for five and six disks. For conjectured extremals for other numbers of disks up to 20, see Zahn.

K. Bezdek, Über einige optimale Konfigurationen von Kreisen, *Ann. Univ. Sci. Budapest. Eötvös Sect. Math.* **27** (1984) 143–151; *MR* **87f**:52020.
J. Molnár, Über eine elementargeometrische Extremalaufgabe, *Mat. Fiz. Lapok.* **49** (1942) 249–253; *MR* **8**, 218.
E. H. Neville, Solutions of numerical functional equations, *Proc. London Math. Soc.* (2) **14** (1915) 308–326.
W. W. Rouse Ball, *Mathematical Recreations and Essays*, 10th ed., Macmillan, New York, 1931, 253–255; 12th ed., University of Toronto Press, Toronto, 1974, 97–99.
S. Verblunsky, On the least number of unit circles which can cover a square, *J. London Math. Soc.* **24** (1949) 164–170; *MR* **11**, 455.
C. T. Zahn, Black box maximization of circular coverage, *J. Res. Nat. Bur. Stand.* B **66** (1962) 181–216.

D4. Packing equal squares in a square.

Let $s(n)$ be the side of the smallest square into which we can pack n unit squares. It is clear that

$$\sqrt{n} \leq s(n) \leq \lceil \sqrt{n} \rceil,$$

with equality when n is a perfect square.

For other values of n, few exact values of $s(n)$ are known: $s(2) = s(3) = 2$, $s(5) = 2 + \frac{1}{2}\sqrt{2} = 2.707...$, and E. Bajmóczy has proved that $s(7) = 3$ [and hence $s(8) = s(9) = 3$]. Figure D2(b) shows that if integers a and b satisfy the inequalities

$$2a^2 - 2a < b^2 < 2a^2,$$

then as many as $n = 2a(a + 1) + b^2$ squares can be packed into a square of side $a + 1 + b/\sqrt{2}$. For example, we get

$(a, b) =$	$(1, 1)$	$(3, 4)$	$(4, 5)$	$(5, 7)$	$(6, 8)$	$(8, 11)$	$(10, 14)$
$n =$	5	40	65	109	148	265	416
$s(n) \leq$	2.7071	6.8285	8.5536	10.4498	12.6569	16.7788	20.8995

Göbel (reference in Introduction to this Chapter) gives some other upper bounds:

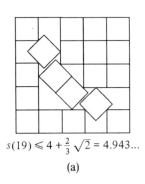
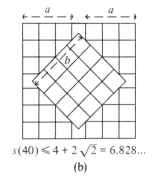

$$s(19) \leqslant 4 + \tfrac{2}{3}\sqrt{2} = 4.943...$$

(a)

$$s(40) \leqslant 4 + 2\sqrt{2} = 6.828...$$

(b)

Figure D2. Efficient packings of a square by (a) 19 equal squares, and (b) 40, and, more generally, $2a(a+1)+b^2$, equal squares.

$n =$	11	18	19	27	28	38	52
$s(n) \leq$	$\frac{5}{2}+\sqrt{2}$	$2+2\sqrt{2}$	$4+2\sqrt{2}/3$	$5+\sqrt{2}/2$	$3+2\sqrt{2}$	$6+\sqrt{2}/2$	$7+\sqrt{2}/2$
	3.9142	4.8284	4.9428	5.7071	5.8284	6.7071	7.7071

Note that the bound for $n = 38$ is better than that given above for $n = 40$.

Erdős & Graham have attacked the problem from a different direction. If n is the maximum number of squares that can be packed into a square of side s, then the waste is $w(s) = s^2 - n$. They have shown that

$$w(s) = O(s^{7/11})$$

while a lower bound has been given by Roth & Vaughan as,

$$w(s) > 10^{-100}(\|s\|s)^{1/2}$$

where $\|s\|$ is the distance from s to the nearest integer.

P. Erdős & R. L. Graham, On packing squares with equal squares, *J. Combin. Theory Ser. A* **19** (1975) 119–123; *MR* **51** #6595.

M. Gardner, Some packing problems that cannot be solved by sitting on the suitcase, *Scientific American* **241** no. 4 (Oct. 1979) 22–26.

K. F. Roth & R. C. Vaughan, Inefficiency in packing squares with unit squares, *J. Combin. Theory Ser A* **24** (1978) 170–186; *MR* **58** #7407.

D5. **Packing unequal rectangles and squares in a square.** Leo Moser noted that $\sum_{n=1}^{\infty} 1/n(n+1) = 1$ and asked if the rectangles $1/n$ by $1/(n+1)$ $(n = 1, 2, ...)$ can be packed into the unit square. Meir & Moser showed that they can be packed into a square of side $31/30$; can this value be reduced?

He also noted that $\sum_{n=2}^{\infty} 1/n^2 = \pi^2/6 - 1 < (5/6)^2$ and asked if the squares of sides $1/2, 1/3, 1/4, ...$ can be packed into a square of side $5/6$. Will they even pack into some *rectangle* of area $\pi^2/6 - 1$?

The squares $s_1^2 \geq s_2^2 \geq \cdots$ with $\sum s_i^2 = 1$ can be placed without overlap in a square of side $s_1 + \sqrt{1 - s_1^2} \leq \sqrt{2}$. Find a better estimate involving more, or all, of the s_i. Owings deals with the case where the s_i are negative powers of 2.

What is the least A such that every set of squares of total area 1 can be packed into some rectangle of area A (the shape of the rectangle can depend on the set)? Kleitman & Krieger showed that the smallest "universal" rectangle that can be packed with *any* set of squares of total area 1, has sides $\sqrt{2}$ and $2/\sqrt{3}$.

Can every set of rectangles, with total area 1, and maximum side 1, be packed into a square of area 2? If they have total area 3 and maximum side 1, will they cover a square of area 1?

The fact that $1^2 + 2^2 + \cdots + 24^2 = 70^2$ suggests that packing these first 24 squares into a square of side 70 might be attempted, but this is impossible. All but the 7^2 can be packed. More generally, what is the smallest square into which squares of integer sides 1 to n will pack? The first few answers are

n	= 1	2	3	4	5	6	7	8	9	10	11	12	13	14	15	16	17	18
square side	1	3	5	7	9	11	13	15	18	21	24	27	30	33	36	39	43	46 or 47
excess area	0	4	11	19	26	30	29	21	39	56	70	79	81	74	56	25	64	

Freedman has suggested the following problem: Given n points in the unit square $0 \le x, y \le 1$, one of which is the origin, can n non-overlapping rectangles be found inside the square and with sides parallel to those of the square, each rectangle having one of the points as its lower left-hand corner, so that the total area is greater than $\frac{1}{2}$? The value $(n + 1)/2n$ is attained when the points are $\{(i/n, i/n) : i = 0, 1, \ldots, n - 1\}$.

What are the analogs of these results in three (or more) dimensions for packing cubes into cubes or rectangular parallelepipeds?

A. Freedman, *Recent Progress in Combinatorics*, Academic Press, New York, 1969.

R. K. Guy, Problem 5, in *The Geometry of Metric and Linear Spaces*, Michigan, 1974, Lecture Notes in Mathematics 490, Springer, Berlin.

D. Kleitman & M. Krieger, Packing squares in rectangles I, *Ann. New York Acad. Sci.* **175** (1970) 253–262; *MR* **41** #9112.

D. Kleitman & M. Kriegar, An optimal bound for two-dimensional bin packing, in *16th Annual Symp. on Foundations of Comput. Sci.*, I.E.E.E. Computer Science, Long Beach, 1975, 163–168; *MR* **54** #11176.

A. Meir & L. Moser, On packing of squares and cubes, *J. Combin. Theory* **5** (1968) 126–134; *MR* **37** #4716.

J. W. Moon & L. Moser, Some packing and covering theorems, *Colloq. Math.* **17** (1967) 103–110; *MR* **35** #6040.

J. Owings, Tiling the unit square with squares and rectangles, *J. Combin. Theory Ser. A* **40** (1985) 156–160; *MR* **87f**:52022.

D6. The Rados' problem on selecting disjoint squares. It is known (see Norlander, Sokolin, and Zalgaller) that, given a finite collection of congruent, similarly-situated squares in the plane that covers a total area A, we may select a nonoverlapping subcollection that covers an area of $\frac{1}{4}A$ or more. T. Rado observed that if the sides of the squares are not necessarily equal (but still parallel) the "greedy algorithm" (repeatedly choose the largest square disjoint from those previously selected) gives a nonoverlapping subset of area $A/9 = A \times 0.111\ldots$, and he asked for a larger bound. R. Rado conjectured that area $\frac{1}{4}A$ was always possible, but could only prove $A \times (4/35) = A \times 0.114\ldots$ Zalgaller improved this to $A/8.6 = A \times 0.116\ldots$ and Mirsky remarked to us that he could obtain $A \times (8/63) = A \times 0.126\ldots$ if only two sizes of square are allowed. However, Ajtai has concocted an example involving hundreds of squares of sides 1 and 2 with no disjoint subset covering area $\frac{1}{4}A$, disproving Rado's conjecture. What then is the largest number c_2 such that there is always a disjoint collection of squares covering area $c_2 A$? From the above remarks $0.116 < c_2 < 0.25$.

Many similar problems may be found in Rado's three papers on the subject. There is, of course, a d-dimensional analog for parallel-sided d-cubes with volume or d-dimensional content replacing area. Improve Rado's bounds $1/(3^d - 7^{-d}) \le c_d \le 2^{-d}$ for the optimal ratio c_d in d dimensions. (That $c_1 = \frac{1}{2}$ is left as an exercise.) Rado considers such problems for disks, centro-symmetric convex sets, etc., as well as for squares.

M. Ajtai, The solution of a problem for T. Rado, *Bull. Acad. Polon. Sci. Sér. Sci. Math. Astr. Phys.* **21** (1973) 61–63; *MR* **47** #7599.

W. Moser & J. Pach, [MP], Problem 42.

G. Norlander, A covering problem, *Nordisk Mat. Tidskr.* **6** (1958) 29–31; *MR* **20** #2670.

R. Rado, Some covering theorems I, *Proc. London Math. Soc.* (2) **51** (1949) 323–364; *MR* **11**, 51.

R. Rado, Some covering theorems II, *Proc. London Math. Soc.* (2) **53** (1951) 243–267; *MR* **13**, 61.

R. Rado, Some covering theorems III, *J. London Math. Soc.* **43** (1968) 127–130; *MR* **40** #1891.

T. Rado, Sur un problème relatif à un théorème de Vitali, *Fund. Math.* **11** (1928) 228–229.

A. Solokin, Concerning a problem of Rado, *C.R. Acad: Sci. U.R.S.S. (N.S)*. **26** (1940) 871–872; *MR* **2**, 256.

V. A. Zalgaller, Remarks on a problem of Rado, *Matem. Prosveskcheric* **5** (1960) 141–148.

D7. **The problem of Tammes.** What is the largest diameter a_n of n equal circles that can be placed on the surface of a unit sphere without overlap? How must the circles be arranged to achieve this maximum, and when is there an essentially unique arrangement? These questions were first raised by the botanist Tammes in 1930 in connection with the distribution of pores on pollen grains.

This problem has an enormous literature; an account is given in the book by Fejes Tóth. The exact solution is known only for $n \le 12$ and $n = 24$. For other values of n, various upper and lower bounds for a_n have been obtained, but many of these could certainly be improved. The table below lists the known values of a_n, given as the angle subtended at the center of the sphere, and indicates the corresponding extremal configurations.

n	a_n	extremal
2	180°	opposite ends of a diameter
3	120°	equilateral triangle in an equatorial plane
4	109°28'	regular tetrahedron
5	90°	regular octahedron less one point (nonunique)
6	90°	regular octahedron
7	77°52'	(unique configuration)
8	74°52'	square anti-prism
9	70°32'	(unique configuration)
10	66°9'	(unique configuration)
11	63°26'	icosahedron less one point (nonunique)
12	63°26'	icosahedron
24	43°41'	snub cube

Clearly $a_n \leq a_{n-1}$ for all n; for which n is there inequality? Certainly for $n = 6$ or 12. Is $a_n = a_{n-1}$ for $n = 24, 48, 60,$ or 120, corresponding to certain highly symmetric arrangements? It is conjectured that equality holds for no other n.

K. Böröczky, The problem of Tammes for $n = 11$, *Studia Sci. Math. Hungar.* **18** (1983) 165–171; *M R* **86j**:52012.

B. W. Clare & D. L. Kepert, The closest packing of equal circles on a sphere, *Proc. Roy. Soc. London Ser. A* **405** (1986) 329–344; *M R* **87i**:52024.

J. H. Conway & N. J. A. Sloane, [CS].

H. S. M. Coxeter, The problem of packing a number of equal nonoverlapping circles on a sphere, *Trans. N.Y. Acad. Sci. Ser II* **24** (1962) 320–331.

L. Danzer, *Endliche Punktmengen auf der 2-Sphäre mit Möglichst Grossen Minimalabstand*, Habilitationsschrift, Göttingen, 1963.

L. Danzer, Finite point-sets on S^2 with minimum distance as large as possible, *Discrete Math.* **60** (1986) 3–66; *M R* **88f**:52014.

L. Fejes Tóth, Über eine Abschätzung des kurzesten Abstandes zweier Punkte eines auf einer Kugelfläche liegenden Punktsystems, *Jber. Deutsch. Math. Verein.* **53** (1943) 66–68; *M R* **8**, 167.

L. Fejes Tóth, [Fej].

L. Fejes Tóth, [Fej'].

M. Gardner, Mathematical games: a numeranalysis by Dr. Matrix of the lunar flight of Apollo 11, *Scientific Amer.* **221** no. 4 (Oct. 1969) 126–130, especially p. 128.

M. Goldberg, Packing of 33 equal circles on a sphere, *Elem. Math.* **18** (1963) 99–100; *M R* **27** #4144.

M. Goldberg, Packing of 18 equal circles on a sphere, *Elem. Math.* **20** (1965) 59–61; *M R* **32** #4605.

M. Goldberg, Packing of 19 equal circles on a sphere, *Elem. Math.* **22** (1967) 108–110.

M. Goldberg, An improved packing of 33 equal circles on a sphere, *Elem. Math.* **22** (1967) 110–111.

M. Goldberg, Axially symmetric packing of equal circles on a sphere, *Ann. Univ. Sci. Budapest. Eötvös Sect. Math.* **10** (1967) 37–48; II *ibid.* **12** (1969) 137–142; *M R* **39** #7510, **41** #9110.

L. Hárs, The Tammes problem for $n = 10$, *Studia Sci. Math. Hungar.* **21** (1986) 439–451; *MR* **89b**:52022.

E. Juçovič, Arrangement of 17, 25 and 33 points on the sphere (Slovak), *Mat.-Fyz. Časopis Sloven. Akad. Vied.* **9** (1959) 173–176; *MR* **23** #A2799.

A. Karabinta & E. Székely, Sur les empilements optimaux de circles congruent sur la sphère, *Ann. Univ. Sci. Budapest. Eötvös Sect. Math.* **16** (1973) 143–154; *MR* **51** #13874.

J. Leech, Equilibrium of sets of particles on a sphere, *Math. Gaz.* **41** (1957) 81–90; *MR* **19**, 165.

J. Leech & T. Tarnai, Arrangements of 22 circles on a sphere, *Ann. Univ. Sci. Budapest. Eötvös Sect. Math.* **31** (1988) 27–37; *MR* **90k**:52028.

E. Makai & T. Tarnai, Morphology of spherical grids, *Acta. Tech. Acad. Sci. Hungar.* **83** (1976) 247–283.

T. W. Melnyk, O. Knop & W. R. Smith, Extremal arrangements of points and unit charges on a sphere: equilibrium configurations revisited, *Canad. J. Chem.* **55** (1977) 1745–1761.

R. M. Robinson, Arrangements of 24 points on a sphere, *Math. Ann.* **144** (1961) 17–48; *MR* **24** #A3565.

R. M. Robinson, Finite sets of points on a sphere with each nearest to five others, *Math. Ann.* **179** (1969) 296–318; *MR* **39** #2068.

H. Rutishauser, Über Punktverteilungen auf der Kugelfläche, *Comment. Math. Helv.* **17** (1945) 327–331; *MR* **7**, 164.

K. Schütte & B. L. van der Waerden, Auf welcher Kugel haben 5, 6, 7, 8 order 9 Punkte mit Mindestabstand eins Platz? *Math. Ann.* **123** (1951) 96–124; *MR* **13**, 61.

J. Strohmajer, Über die Verteilung von Punkten auf der Kugel, *Ann. Univ. Sci. Budapest. Eötvös Sect. Math.* **6** (1963) 49–53; *MR* **29** #3958.

E. Székely, Sur le problème de Tammes, *Ann. Univ. Sci. Budapest. Eötvös Sect. Math.* **17** (1974) 157–175; *MR* **52** #4141.

P. M. L. Tammes, on the origin of number and arrangement of the places of exit on the surface of pollen grains, *Recueil Travaux Botaniques Néerlandais.* **27** (1930) 1–84.

T. Tarnai, Packing of 180 equal circles on a sphere, *Elem. Math.* **38** (1983) 119–122; *MR* **85** c:52022; correction, *ibid.* **39** (1984) 129; *MR* **87f**:52023.

T. Tarnai, Note on packing of 19 equal circles on a sphere, *Elem. Math.* **39** (1984) 25–27; *MR* **87b**:52035.

T. Tarnai, Multisymmetric packing of equal circles on a sphere, *Ann. Univ. Sci. Budapest. Eötvös Sect. Math.* **27** (1984) 199–204; *MR* **87c**:52026.

T. Tarnai & Zs. Gáspár, Improved packing of equal circles on a sphere and rigidity of its graph, *Math. Proc. Cambridge Philos. Soc.* **93** (1983) 191–218; *MR* **85b**:52014.

T. Tarnai & Zs. Gáspár, Multisymmetric close packings of equal spheres on the spherical surface, *Acta Cryst. Sect. A* **43** (1987) 612–616; *MR* **89d**:52031.

B. L. van der Waerden, Punkte auf der Kugel. Drei Zusätze, *Math. Ann.* **125** (1952) 213–222; Berichtigung, *ibid.* **152** (1963) 94; *MR* **14**, 401.

D8. **Covering the sphere with circular caps.** What is the smallest radius r_n of n equal circles (or spherical caps) that can cover the unit sphere? This problem is the "covering" analog of the Tammes packing problem (see D7). The values of r_n and the corresponding arrangements of circles have been found for $n \leq 7$ and $n = 10$, 12, and 14 and estimates have been obtained for other values of n.

G. Fejes Tóth, Kreisüberdeckungen der Sphäre, *Studia Sci. Math. Hungar.* **4** (1969) 225–247; *MR* **42** #6724.

L. Fejes Tóth, Covering a spherical surface with equal spherical caps, *Mat. Fiz. Lapok.* **50** (1943) 40–46; *MR* **8**, 219.

L. Fejes Tóth, [Fej].

T. Tarnai & Zs Gáspár, Covering the sphere with equal circles, in [BF], 545–550.

T. Tarnai & Zs Gáspár, Covering the sphere with 11 equal circles, *Elem. Math.* **41** (1986) 35–38; *MR* **88c**:52014.

D9. Variations on the penny-packing problem. Arrange n non-overlapping unit disks ("pennies") in $\mathbb{R}^2$ or balls in $\mathbb{R}^d (d \geq 2)$ so as to minimize the area or volume of their convex hull. For $d = 2$, the convex hull tends to be "as hexagonal as possible [see Figure D3(a)]; this is made precise by Wegner for many values of n, including all $n \leq 120$, and also for $n = 3k^2 + 3k + 1$, corresponding to exact hexagonal arrangements.

The situation is somewhat different in $\mathbb{R}^3$. For $n \leq 56$, the best arrangements are **sausages**, that is, with the centers of the balls along a straight line, so that neighboring balls touch [see Figure D3(b)]. However for larger n, the arrangements whose convex hulls have minimum volume tend to be clusters. In $\mathbb{R}^4$, the same phenomenum occurs with sausages providing the best configurations for up to somewhere between 50,000 and 100,000 balls. For dimensions $d \geq 5$, Fejes Tóth makes the "sausage conjecture," that the minimum volume of the convex hull of n disjoint balls is achieved by a sausage for all n. The underlying idea is that any other packing leaves so much space between the balls that it cannot be optimal.

Some special cases of this conjecture have been proved, generally under a fairly restrictive assumption that the centers of the spheres all lie in a hyperplane or a lower-dimensional plane. A survey of known results is given by

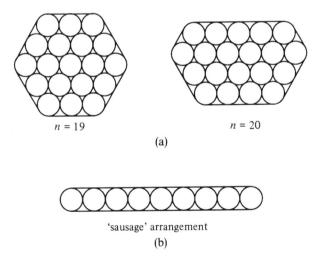

$n = 19$ $\qquad\qquad\qquad$ $n = 20$

(a)

'sausage' arrangement

(b)

Figure D3. (a) Packing n disks so that their convex hull has least area for $n = 19$ and 20. The optimal packing tends to be "fairly hexagonal". (b) The "sausage" packing, believed to lead to the optimal packings of d-dimensional spheres in R^d for $d \geq 5$.

Gritzmann & Wills. Betke, Gritzmann & Wills show that the conjecture is true whenever $\dim S \leq 3$ or $\dim S \leq \frac{7}{12}(d-1)$, where $\dim S$ is the dimension of the convex hull of the set S of centers. Betke & Gritzmann treat the cases when $\dim S \leq 9$ and $d > 1 + \dim S$ and also if the centers lie on certain lattices, and Kleinschmidt, Pachner & Wills consider $\dim S = 2$ and where $d \geq 75(\dim S)^3$. However this is a long way from the sausage conjecture in its general form.

Gritzmann proved that the volume of the convex hull of a packing of balls in $\mathbb{R}^d$ is always greater than $1/[2 + \sqrt{2} + 2(d-1)^{-1/2}]$ times the volume of the convex hull of the sausage packing. Can this figure be increased, even if not to 1?

For similar problems involving packing n disks or balls into a convex set so that the perimeter or surface area, or alternatively the diameter, is minimum, it seems likely that the optimum shape is roughly circular or spherical if n is large. Recently, Graham & Sloane have sought configurations minimizing the moment of inertia about the centroid.

U. Betke & P. Gritzmann, Über L. Fejes Tóths Wurstvermutung in kleinen Dimensionen, *Acta Math. Hungar.* **43** (1984) 299–307; *MR* **85g**:52006.

U. Betke & P. Gritzmann, An application of valuation theory to two problems in discrete geometry, *Discrete. Math.* **58** (1986) 81–85.

U. Betke, P. Gritzmann & J. M. Wills, Slices of L. Fejes Tóth's sausage conjecture, *Mathematika* **29** (1982) 194–201; *MR* **84m**:52017.

G. Fejes Tóth, P. Gritzmann & J. M. Wills, Finite sphere packing and sphere covering, *Discrete Comput. Geom.* **4** (1989) 19–40; *MR* **89k**:52016.

L. Fejes Tóth, Research Problem 13, *Periodica Math. Hungar.* **6** (1975) 197–199.

R. L. Graham & N. J. A. Sloane, Penny packing and two-dimensional codes, *Discrete Comput. Geom.* **5** (1990) 1–11; *MR* **90j**:52013.

P. Gritzmann, Finite packing of equal balls, *J. London Math. Soc.* (2) **33** (1986) 543–553; *MR* **87k**:52035.

P. Gritzmann & J. M. Wills, Finite packing and covering, *Studia Sci. Math. Hungar.* **21** (1986) 149–162; *MR* **88h**:52022.

P. Kleinschmidt, U. Pachner & J. M. Wills, On L. Fejes Tóth's "sausage conjecture," *Israel J. Math.* **47** (1984) 216–226; *MR* **85d**:52008.

G. Wegner, Über endliche Kreispackungen in der Ebene, *Studia Sci. Math. Hungar.* **21** (1986) 1–28; *MR* **88i**:52024.

J. M. Wills, On the density of finite packings, *Acta. Math. Hungar.* **46** (1985) 205–210; *MR* **87e**:52025.

D10.   Packing balls in space. The "most dense" or "most efficient" packing of the plane by disks of equal radius is given by the arrangement with the centers of the disks at the points of a regular triangular lattice. To be more precise, we defined the **density** of a packing as

$$\lim_{r \to \infty} (\text{area of disks contained in } B_r)/(\text{area of } B_r),$$

where B_r is the disk center the origin and radius r. Then the maximum possible density of any packing by equal disks is $\pi/2\sqrt{3} = 0.9069\ldots$, the value achieved by the triangular lattice packing. For proofs of this fact, which dates back at least to Gauss, see the books of Rogers or Meschkowski.

The 3-dimensional analog of this problem is one of the famous unsolved problems in mathematics. We define the **density** of a packing of balls in space as

$$\lim_{r \to \infty} (\text{volume of balls contained in } B_r)/(\text{volume of } B_r),$$

where B_r is now the ball of radius r and center the origin. There is a natural way to pack balls (of radius $1/\sqrt{2}$), layer upon layer, so that the centers are at the points with integer coordinates (x, y, z) for which the sum $x + y + z$ is even. This packing has density $\pi/3\sqrt{2} = 0.7404\ldots$. To quote Rogers, "Many mathematicians believe and all physicists know" that this is the greatest density possible. However, this has not been proved for general packings, though Gauss showed that it is the maximum density among lattice packings, where the centers of the balls are at the points of some lattice. One of the difficulties is that it is possible to have extensive partial packings for which the local density is unexpectedly high.

In 1958, Rogers showed that no packing can have a density greater than $0.77963\ldots$. Slight improvements have been obtained very recently, by Lindsey $(0.77844\ldots)$, Muder $(0.77836\ldots)$, and Lindsey $(0.7736\ldots)$.

Considerable progress has been made recently in obtaining bounds for packing densities of balls in higher dimensions. The book by Conway & Sloane provides an excellent and detailed account of this work, and includes the best known estimates for packing densities in dimensions up to more than 1,000,000. Very recently, Noam Elkies has used elliptic curves to obtain new records for many dimensions between 80 and 1000.

J. H. Conway & N. J. A. Sloane, [CS].
N. Elkies, Elliptic curves and lattices, to appear.
J. H. Lindsey II, Sphere packing in R^3, *Mathematika* **33** (1986) 137–147; *MR* **87m**:52026.
H. Meschkowski, [Mes], Chapters 2 and 3.
D. J. Muder, Putting the best face on a Voronoi polyhedron, *Proc. London Math. Soc.* (3) **56** (1988) 329–348.
C. A. Rogers, [Rog].
C. A. Rogers, The packing of equal spheres, *Proc. London Math. Soc.* (3) **8** (1958) 609–620; *MR* **21** #847.
N. J. A. Sloane, The packing of spheres, *Sci. American* **250** (Jan. 1984) 116–125.

D11. **Packing and covering with congruent convex sets.** We define the **density** of a collection of convex sets $\{K_i\}$ in the plane as

$$\lim_{r \to \infty} (\pi r^2)^{-1} \sum_i A(B_r \cap K_i),$$

assuming that this limit exists, where B_r is the disk of radius r and center the origin, and where A denotes area.

The **packing density** $\delta(K)$ of a convex set K is the greatest possible density of a packing of the plane by congruent copies of K (disjoint except for boundary points). Thus, for example, $\delta(K) = 1$ if K is any triangle or quadrilateral.

What is the least value of $\delta(K)$ among (proper, bounded) convex sets K, and for which K is this minimum attained? In other words, which sets K can be packed least efficiently? Chakerian & Lange showed that every convex set K is contained in a quadrilateral of $\sqrt{2}$ times its area; since congruent copies of any quadrilateral tile the plane, it follows that $\delta(K) \geq 1/\sqrt{2}$ for any K. Rather more complicated methods of packing have led to improved bounds, the current best estimate being $\delta(K) \geq \sqrt{3}/2 = 0.8660\ldots$ due to Kuperberg & Kuperberg. Can this be increased, perhaps even to $\pi/2\sqrt{3} = 0.9069\ldots$, the packing density of a circular disk?

There are numerous variations on this problem. For packings by *translates* of a convex set K, it is known that $\delta(K) \geq \frac{2}{3}$ with equality only when K is a triangle. But what is the least value of $\delta(K)$ when K is a centro-symmetric set, either for packings by congruent copies of K, or alternatively by translates of K?

Dually, the **covering density** $\theta(K)$ is the smallest possible density of any arrangement of congruent copies of K that cover the plane. What is the least upper bound of $\theta(K)$ for convex sets K? Kuperberg has shown that $\theta(K) \leq 8(2\sqrt{3} - 3)/3 = 1.237\ldots$ It is conjectured that we always have $\theta(K) \leq 2\pi/\sqrt{27} = 1.2092\ldots$, the value for the circular disk, and the maximum value among centro-symmetric sets. If only translates of K are allowed, then $\theta(K) \leq \frac{3}{2}$, with equality for a triangle.

There are obvious higher-dimensional analogs to these problems, where density is defined by

$$\lim_{r \to \infty} V(B_r)^{-1} \sum_i V(B_r \cap K_i)$$

with B_r the ball of radius r in $\mathbb{R}^d$ and V denoting d-dimensional volume. However, little is known even in $\mathbb{R}^3$ and only very rough bounds for packing and covering densities have been established. Here, calculation or estimation of packing or covering densities of specific sets presents challenging enough problems. For example, find good estimates for the packing density of the regular tetrahedron or octahedron in $\mathbb{R}^3$. The case of the sphere is discussed in D10.

The reader is urged to consult Fejes Tóth or Rogers for further details.

G. D. Chakerian & L. H. Lange, Geometric extremum problems, *Math. Mag.* **44** (1971) 57–69; *MR* **46** #794.

L. Fejes Tóth, [Fej].

R. Kershner, The number of circles covering a set, *Amer. J. Math.* **61** (1939) 665–671.

W. Kuperburg, Packing convex bodies in the plane with density greater than 3/4, *Geom. Dedicata* **13** (1982) 149–155; *MR* **84d**:52018.

W. Kuperburg, An inequality linking packing and covering densities of plane convex bodies, *Geom. Dedicata* **23** (1987) 59–66; *MR* **88h**:52023.

W. Kuperburg, On packing the plane with congruent copies of a convex body, in [BF], 317–329; *MR* **88j**:52038.

W. Kuperburg, Covering the plane with congruent copies of a convex body, *Bull. London Math. Soc.* **21** (1989) 82–86; *MR* **89m**:52024.

G. Kuperburg & W. Kuperburg, Double-lattice packings of convex bodies in the plane, *Discrete Comput. Geom.*, to appear.

C. A. Rogers, [Rog].

E. Sas, Über eine Extremumeigenschaft der Ellipsen, *Compositio Math.* **6** (1939) 468–470.

D12. Kissing numbers of convex sets.

The **kissing number** of a convex set K in $\mathbb{R}^d$ is the greatest number of disjoint congruent copies of K that can be arranged around K so that they all touch K. In the plane, it is well-known that the disk has a kissing number of 6 (arrange 7 pennies in a hexagonal pattern!) but the Reuleaux triangle has a kissing number of seven. Fejes Tóth asks whether there exists a plane set of constant width with a kissing number of 8 and also whether it is possible for 7 disjoint copies of a set of constant width K to touch K if we insist that they are *translates* of K.

The kissing number of the ball in $\mathbb{R}^3$ is 12, but in $\mathbb{R}^4$ it is uncertain whether it is 24 or 25. For larger d, the kissing number of the ball in $\mathbb{R}^d$ is only known exactly when $d = 8$ (240) and $d = 24$ (196560); these arrangements correspond to certain highly symmetrical lattices in $\mathbb{R}^8$ and $\mathbb{R}^{24}$. A table of known bounds for $d \leq 24$ is given by Odlyzko & Sloane and in Conway & Sloane, page 23. How do the kissing numbers of balls behave as $d \to \infty$? Wyner showed (non-constructively) that arrangements exist with a ball touching $2^{0.2075d[1+O(1)]}$ congruent ones, and Kabatiansky & Levenshtein showed that at most $2^{0.401d[1+O(1)]}$ is possible. What is the exact exponent?

J. H. Conway & N. J. A. Sloane, [CS].

H. S. M. Coxeter, An upper bound for the number of equal non-overlapping spheres that can touch another of the same size, in [Kle], 53–71; *MR* **29** #1581.

L. Fejes Tóth, On the number of equal discs that can touch another of the same kind, *Studia Sci. Math. Hungar.* **2** (1967) 363–367; *MR* **36** #4440.

G. A. Kabatjansky & V. I. Levenšteĭn, Bounds for packings of the sphere and in space, Prob. Information Trans. **14** (1978) 1–17.

J. Leech, The problem of the thirteen spheres, *Math. Gaz.* **40** (1956) 22–23; *MR* **17**, 888.

A. M. Odlyzko & N. J. A. Sloane, New bounds on the number of unit spheres that can touch a unit sphere in n dimensions, *J. Combin. Theory Ser. A* **26** (1979) 210–214; *MR* **81d**:52010.

K. Schütte & B. L. van der Waerden, *Das Problem der dreizehn Kugeln*, *Math. Ann.* **125** (1953) 325–334; *MR* **14**, 787.

A. D. Wyner, Capabilities of bounded discrepancy decoding, *AT&T Tech. J.* **44** (1965) 1061–1122.

D13. Variations on Bang's plank theorem.

A **plank** or **slat** is the (closed) region between two parallel lines in $\mathbb{R}^2$, planes in $\mathbb{R}^3$, and hyperplanes in higher dimensions. Tarski conjectured that if a convex body K of minimal width w is convered by a set of k planks of widths $p_1, \ldots, p_k$, then $\sum_1^k p_i \geq w$, equality of course being achieved by using a single plank of width w, [see Figure D4(a)]. A proof was provided by Bang, and this was refined by Fenchel and then by Bognar. In fact, Bang proved the stronger result that $\sum_1^k p_i/l_i \geq 1$, where l_i is

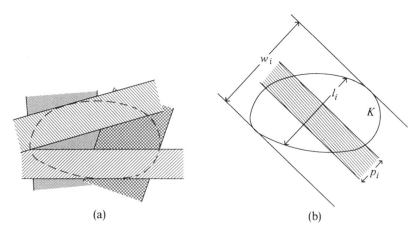

Figure D4.   (a) Cover of a convex set by four planks. (b) Quantities relating to the relative width version of the plank problem.

the length of the longest chord of K perpendicular to the ith plank, [see Figure D4(b)]. The outstanding unsolved problem is the affine-invariant version of the result: if w_i is the width of K in the direction perpendicular to the ith plank, is it true that $\sum_1^k p_i/w_i \geq 1$?

Some proofs of this have been published but not everyone is convinced by them. It is certainly true if there are just two planks. Very recently, Ball has shown that it is true if K is centro-symmetric about the origin. Rogers suggests attacking the case where K is a regular d-dimensional simplex when many different coverings seem to yield equality.

Alexander pointed out that the affine-invariant conjecture for a convex set K is equivalent to the following property: given any n lines cutting K, there exists a homothetic (i.e., similar and similarly situated) copy of K with scale factor $1/(n + 1)$ contained in K that does not cut any of the lines.

Fejes Tóth posed the following related problem. A **zone of width** w on the surface of a sphere is defined to be the parallel domain consisting of points within angular distance $w/2$ of a great circle. Prove (or disprove) that the total width of any collection of zones covering the sphere is at least π.

Another variant of Tarski's problem is to consider multiple coverings. If we insist that each point of K be covered by at least k planks, what is the largest constant $c(k)$ such that $\sum p_i \geq c(k)w$? For the equilateral triangle $c(k) < k$ if $k \geq 2$. Perhaps most interesting is the limiting case as $k \to \infty$. The problem becomes that of minimizing $\sum_1^k \int f_i(t)\, dt$ where the f_i are nonnegative functions and θ_i are unit vectors such that $\sum_1^k f_i(\mathbf{x} \cdot \theta_i) \geq 1$ for all $\mathbf{x} \in K$. See the paper of Falconer for further details.

Further variations involve covering a convex set by parallel translates of a given collection of slabs. It is not even known whether the 3-dimensional unit cube can be covered by translates of any countable collection of slabs with $\sum_1^\infty p_i = \infty$. See Section 9 of Groemer's article for further problems of this type.

R. Alexander, A problem about lines and ovals, *Amer. Math. Monthly* **75** (1968) 484–487.

K. Ball, The plank problem in general normed spaces, to appear.

T. Bang, A solution of the "plank problem," *Proc. Amer. Math. Soc.* **2** (1951) 990–993.

T. Bang, Some remarks on the union of convex bodies, in *Proceedings, Tolfte Skandinaviska Matematikerkongressen*, Lunds Universitets Math. Inst., Lund, 1953, 5–11.

N. Bognar, On Fenchel's solution of the plank problem, *Acta Math. Acad. Sci. Hungar.* **12** (1961) 269–270; *MR* **24** #A2893.

K. J. Falconer, Function space topologies defined by sectional integrals and applications to an extremal problem, *Math. Proc. Cambridge Philos. Soc.* **87** (1980) 81–96; *MR* **82d**:52010.

L. Fejes Tóth, [Fej'].

W. Fenchel, On Th. Bang's solution of the plank problem, *Mat. Tidsskr.* B **1951** (1951) 49–51; *MR* **13**, 863.

R. J. Gardner, Relative width measures and the plank problem, *Pacific J. Math.* **135** (1988) 299–312; *MR* **89j**:52009.

H. Groemer, Coverings and packings by sequences of convex sets, in [GLMP], 262–278; *MR* **87a**:52016.

W. O. J. Moser, On the relative widths of coverings by convex bodies, *Canad. Math. Bull.* **1** (1958) 154, 168, 174; *MR* **23** #A563.

C. A. Rogers, Some problems in the geometry of convex bodies, in [DGS], 279–284; *MR* **84b**:52001.

A. Tarski, Further remarks about the degree of equivalence of polygons, *Odbitka z Parametru* **2** (1932) 310–314.

D14. **Borsuk's conjecture.** One of the most famous unsolved problems of geometry is Borsuk's conjecture: Every subset K of $\mathbb{R}^d$ of unit diameter can be covered by $d + 1$ sets each of diameter strictly less than 1. (In fact Borsuk was careful never to express this as a conjecture, but merely as a question.) This is certainly false for any number less than $d + 1$, since if the surface of a sphere in $\mathbb{R}^d$ is divided into d closed sets, then at least one of the sets contains a pair of antipodal points.

Any set is contained in a convex body, indeed in a convex body of constant width, of the same diameter, so a proof when K is convex or even of constant width, would suffice.

In the case where $d = 2$, the conjecture follows from the fact that any set of diameter one is contained in a regular hexagon of width 1 which splits into three congruent parts of diameter $\sqrt{3}/2$ (see Figure D5).

In three dimensions, the conjecture was verified by Perkal and also by Eggleston using rather complicated arguments. However Grünbaum and Heppes independently showed that any set of diameter 1 is contained in a regular octahedron of width 1 with three of its vertices cut off by planes perpendicular to the main diagonals at a distance 1/2 from the center, leading to a dissection into four parts of diameter less than 0.9887.... This value is the smallest known at present, though Gale conjectures that there is always a dissection into four parts of diameter at most $\sqrt{(3 + \sqrt{3})/6} = 0.888...$.

In higher dimensions, the question is still open, except in certain special cases: when K is a convex set with every boundary point lying in exactly one hyperplane (Hadwiger, Eggleston); when K is centro-symmetric (Riesling);

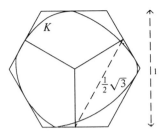

Figure D5. A proof of Borsuk's conjecture in the plane. A set K of diameter 1 has a circumscribing regular hexagon of side 1 which may be divided into three parts of diameter $\frac{1}{2}\sqrt{3} < 1$.

when K is invariant under the symmetry group of the regular d-dimensional simplex (Rogers).

Melzak lists three (presumably rather difficult) problems, an affirmative answer to any of which would imply the truth of Borsuk's conjecture (see A27).

Borsuk's conjecture would follow if it could be shown that any *finite* set K in $\mathbb{R}^d$ can be split into $d + 1$ sets of diameter at most a, for some fixed $a < 1$. Is it even true that such a K can be split into $d + 1$ sets of *some* diameter (dependent on K) less than 1? Larman has suggested investigating the problem when K is a **two-distance set** [so that if $\mathbf{x}, \mathbf{y} \in K$ then $|\mathbf{x} - \mathbf{y}| = 1$ or $|\mathbf{x} - \mathbf{y}| = c$ for some $c < 1$ (see F3)]. This very special case seems to contain many of the difficulties intrinsic to the general problem.

One can at least try to obtain good estimates for $k(d)$, the smallest number such that any $K \subset \mathbb{R}^d$ of diameter 1 can be partitioned into $k(d)$ pieces of diameter strictly less than 1. Danzer proved that $k(d) < 3^{-1/2}(d + 2)^{3/2}(2 + \sqrt{2})^{(d-1)/2}$, and more recently Lassak obtained $k(d) \le 2^{d-1} + 1$ and Schramm $k(d) \le 5d^{3/2}(4 + \ln d)(3/2)^{d/2}$.

More generally, let $D(K, n)$ be the smallest number such that the convex set K of diameter 1 can be split into n subsets of diameters at most $D(K, n)$. Thus Borsuk's conjecture is that $D(K, d + 1) < 1$ for all K in $\mathbb{R}^d$. Estimate $D(K, n)$ for various convex bodies K. It was shown by Lenz and others that $D(K, 3) \le \frac{1}{2}\sqrt{3}$, $D(K, 4) \le \frac{1}{2}\sqrt{2}$, and $D(K, 7) \le \frac{1}{2}$, where K is a convex subset of the plane of diameter 1, with equality for the disk. However, $D(\text{regular } 11\text{-gon}, 6) = 0.5051 \cdots > \frac{1}{2} = D(\text{disk}, 6)$. What is the maximum value of $D(K, n)$ for K in $\mathbb{R}^2$ or $\mathbb{R}^3$ for other small n? Is it true that, for any K in $\mathbb{R}^3$, we have $D(K, 4) \le D(\text{ball in } \mathbb{R}^3, 4)$ $[= \sqrt{(3 + \sqrt{3})/6}]$? More generally, what is the value of $D(\text{ball in } \mathbb{R}^d, d + 1)$ for $d \ge 4$, and is this the maximum of $D(K, d + 1)$ for any K in $\mathbb{R}^d$? [Of course, $D(\text{ball in } \mathbb{R}^d, d) = 1$ for all d.]

More detailed accounts of progress on Borsuk's conjecture and related topics are given in Grünbaum's survey and in the tract by Boltjanskii & Gohberg.

V. Boltjanskii & I. Gohberg, *Results and Problems in Combinatorial Geometry*, Cambridge University Press, Cambridge, 1985; *MR* 87c:52002.

K. Borsuk, Drei Sätze über die *n*-dimensionale euklidische Sphäre, *Fund. Math.* **20** (1933) 177–190.

L. Danzer, Über Durchschnitteigenschaften *n*-dimensionalen Kugelfamilien, *J. Reine Angew. Math.* **209** (1961) 181–203; *MR* **25** #5453.

B. V. Dekster, Borsuk's covering for blunt bodies, *Arch. Math. (Basel)* **51** (1988) 87–91; *MR* **89k**:52015.

H. G. Eggleston, Covering a three-dimensional set with sets of smaller diameter, *J. London Math. Soc.* **30** (1955) 11–24; *MR* **16**, 734.

H. G. Eggleston, *Problems in Euclidean Space: Application of Convexity*, Pergamon, New York, 1957; *MR* **23** #A3328.

D. Gale, On inscribing *n*-dimensional sets in a regular *n*-simplex, *Proc. Amer. Math. Soc.* **4** (1953) 222–225; *MR* **14**, 787.

B. Grünbaum, A simple proof of Borsuk's conjecture in three dimensions, *Proc. Cambridge Philos. Soc.* **53** (1957) 776–778; *MR* **19**, 763.

B. Grünbaum, Borsuk's problem and related questions, in [Kle], 271–284; *MR* **27** #4134.

R. K. Guy & J. L. Selfridge, Optimal coverings of squares, *Colloq. Math. Soc. János Bolyai* **10** (1973) 745–799; *MR* **51** #13873.

H. Hadwiger, Überdeckung einer Menge durch Mengen kleineren Durchmesse, *Comment. Math. Helv.* **18** (1945) 73–75; *MR* **7**, 215.

H. Hadwiger, Von der Zerlegung der Kugel in Kleinere Teile, *Gaz. Math. Lisboa* **15** (1954) 1–3; *MR* **15**, 982.

A. Heppes, On the partitioning of three-dimensional point sets into sets of smaller diameter (Hungarian), *Magyar Tud. Akad. Mat. Fiz. Oszt. Kösl.* **7** (1957) 413–416; *MR* **20** #1952.

D. G. Larman, Problem 41, [TW].

H. Lenz, Zerlegung ebener Bereiche in konvexe Zellen von möglichst kleinen Durchmesser, *Jber. Deutsch. Math. Verein* **58** (1956) 87–97; *MR* **18**, 817.

H. Lenz, Über die Bedeckung ebener Punktmengen durch solche kleineren Durchmessers, *Arch. Math.* **7** (1956) 34–40; *MR* **17**, 888.

M. Lassak, An estimate concerning Borsuk partition problem, *Bull. Acad. Polon. Sci. Sér. Sci. Math.* **30** (1982) 449–451; *MR* **84j**:52014.

Z. A. Melzak, More problems connected with convexity, Problems 7–9, *Canad. Math. Bull.* **11** (1968) 482–494; *MR* **38** #3767.

W. Moser & J. Pach, Problems 28 and 29, [MP].

J. Perkal, Sur la subdivision des ensembles en parties de diamètre intérieure, *Colloq. Math.* **1** (1947) 45.

A. S. Riesling, Borsuk's problem in three-dimensional spaces of constant curvature, *Ukr. Geom. Sbornik* **11** (1971) 78–83.

C. A. Rogers, Symmetrical sets of constant width and their partitions, *Mathematika* **18** (1971) 105–111; *MR* **44** #7432.

O. Schramm, Illuminating sets of constant width, *Mathematika* **35** (1988) 180–189.

D15. Universal covers. A subset of the plane is called a **universal cover** if it can be moved to cover any plane set of diameter 1 (rotations allowed). A simple example of a universal cover is the Reuleaux triangle of width 1 with a semi-circle adjoined to one side, [Figure D6(a)] having area 1.008... (see Eggleston).

An old and well-known problem of Lebesgue is to find the convex universal cover that is of least area. Years ago Pál pointed out that the regular hexagon of width 1 and area $\sqrt{3}/2 = 0.866...$ was a universal cover, and

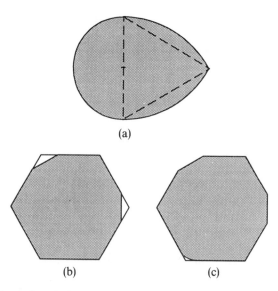

Figure D6. Examples of plane universal covers. (a) Reuleux triangle with adjoined semicircle, area 1.008.... (b) Truncated regular hexagon, area 0.845.... (c) Further reduction of (b), area 0.8441....

also that this could be reduced to a universal (irregular) octagon of area $2 - \frac{2}{3}\sqrt{3} = 0.845...$ by removing two corners of the hexagon [Figure D6(b)]. The smallest examples known seem to be those of Sprague and of Duff with area 0.8441... [see Figure D6(c)]. Hansen has a cover of area 0.8430... for all centro-symmetric sets which he conjectures is universal. Pál points out that any convex universal cover must have an area at least 0.826321..., the area of the smallest convex set containing both a circle and an equilateral triangle of unit diameter.

An outstanding problem is whether the convex universal cover of minimum area is unique. One can also try to deduce qualitative properties of the smallest universal covers without necessarily identifying the sets, for example, do they have smooth boundaries?

In the same vein what are the convex universal covers of least perimeter length? Edwards asks for sets of small area or perimeter that cover all convex sets with perimeter of length 1.

We can ask for small universal covers of particular types, for example, convex sets that are centro-symmetric or of constant width. Chakerian & Logothetti have shown that, for $n \geq 4$, the smallest convex n-gon that is a universal cover, coincides with the smallest regular n-gon that covers the Reuleaux triangle of width 1.

Of course, similar questions can be asked for non-convex universal covers. Kovalev has shown that covers of minimum area must be star-shaped, and obtains a Lipschitz condition on the boundary curves. But can the smallest such sets be identified?

Very little is known about universal covers in three dimensions, where analogous problems arise. A question due to Grünbaum is whether there is a number r such that every irreducible convex universal cover has a diameter of less than r. (A universal cover is **irreducible** if no proper subset is universal.) This is known to be true in two dimensions.

A common approach to Borsuk's conjecture (see D14) is to partition universal covers into subsets of a diameter of less than 1, giving a corresponding dissection of any convex subset of the cover.

More detailed surveys of these problems are given by Grünbaum and by Meschkowski.

D. Chakerian & D. Logothetti, Minimal regular polygons serving as universal covers in $\mathbb{R}^2$, *Geom. Dedicata* **26** (1988) 281–297; *MR* **89f**:52035.

G. F. D. Duff, A smaller universal cover for sets of unit diameter, *Roy. Soc. Canad. Math. Report* **2** (1980) 37–42.

A. Edwards, Professor Besicovitch's minimal problem—a challenge. *Eureka* **20** (1957) 26–27.

H. G. Eggleston, Minimal universal covers in E^n, *Israel J. Math.* **1** (1963) 149–155; *MR* **28** #4432.

B. Grünbaum, Borsuk's problem and related questions, in [Kle], 271–284; *MR* **27** #4134.

H. C. Hansen, Towards the minimal universal cover, *Normat.* **29** (1981) 115–119, 148; *MR* **83g**:52011.

M. D. Kovalev, The smallest Lebesgue covering exists, *Mat. Zametki* **40** (1986) 401–406, 430 (translation *Math. Notes* **40** (1986) 736–739); *MR* **88h**:52008.

H. Meschkowski, [Mes], Chapter 5.

W. Moser & J. Pach, Problem 30, [MP].

J. Pál, Über ein elementaires Variationesproblem, *Danske Vid. Selsk. Math. Fys. Medd.* **3** (1920), no. 2.

R. Sprague, Über ein elementaires Variationesproblem, *Mat. Tidschrift* (1936) 6–99.

B. Weissbach, Polyhedral covers, in [BF], 639–646.

D16. **Universal covers for several sets.** A consequence of a result of Falconer is that every plane curve C of unit length is contained in a semi-circle of radius at most $1/\pi$. Lutwak observed that this implies that any two plane convex sets each with a perimeter length of at most 1 may be packed into a disk of radius $1/\pi$ without overlapping. In other words, any disk that can accommodate two disks of unit perimeter can accommodate *any* two convex sets of unit perimeter. Lutwak asks whether this is true for larger numbers of convex sets: does any disk that accommodates k disks of unit perimeter necessarily accommodate any k convex sets of unit perimeter? If not, what is the smallest disk that does so? Do results of this nature hold if the perimeters of the sets are not all equal?

Lutwak points out that for the 3-dimensional analog of the problem we should ask whether every sphere that accommodates two balls of unit width will necessarily accommodate any two convex bodies of unit mean width.

Clearly one can postulate a wide variety of problems of this nature. How large does a convex set of a particular type need to be in order to contain any k convex sets of given "sizes"?

It is natural to think of these problems in relation to the worm problem (see D18) and Lesbegue's covering problem (see D15).

K. J. Falconer, A characterization of plane curves of constant width, *J. London Math. Soc.* (2) **16** (1977) 536–538; *MR* **57** #1272.

E. Lutwak, On packing curves into circles, in [KB], 107–111; *MR* **83g**:52012.

D17. Hadwiger's covering conjecture.

Let $h(K)$ be the smallest number such that a compact convex body K in $\mathbb{R}^d$ can be covered by $h(K)$ translates of cK for some $c < 1$ ($cK = \{cx : x \in K\}$). Thus we seek coverings of K by smaller homothetic (similar and similarly situated) copies of K. Hadwiger conjectured that $h(K) \leq 2^d$ with equality, if and only if K is a parallelepiped. Levi has proved this for $d = 2$. Little seems to be known in general for higher-dimensional K, but if K has a smooth boundary then $h(K) \leq d + 1$, if K is centro-symmetric then $h(K) \leq 2^d d(\ln d + \ln \ln d + 5)$, and if K is of constant width then $h(K) \leq 5(3/2)^{d/2} d^{3/2}(\ln d + 4)$ seem to be the best estimates. Recently Lassak has shown that $h(K) \leq 8$ if K is a 3-dimensional centro-symmetric body. The article by Grünbaum and the tract by Boltyanskii & Gohberg on the closely related Borsuk conjecture, includes detailed discussions of the problem and further conjectures.

Remarkably, the following "illumination problem" is an equivalent one: If x is a "light source" exterior to K, a point y on the surface of K is illuminated from x if the line through x and y cuts the interior of K with the line segment (x, y) disjoint from K. Boltyanskii showed that the "Hadwiger number" $h(K)$ equals the least number of exterior sources required to illuminate the entire surface of K.

Also of interest is the more general problem: given n, what is the least value of c such that any convex set K may be covered by n translates of the set cK? It is obvious, by considering areas that no plane convex body K can be coverd by 4 translates of cK if $c < \frac{1}{2}$. Is the same true for 5 or even 6 translates? Other known and conjectured estimates for $n \leq 9$ are given in the survey by Lassak.

V. G. Boltyanskii, Problem of illumination of the boundary of a convex body, *Izv. Mold. Fil. Akad. Nauk. SSSR* **10** (76) (1960) 79–86.

V. G. Boltyanskii & P. S. Soltan, *Combinatorial Geometry of Various Classes of Convex Sets*, Shtinista Kishimer, 1978.

V. G. Boltyanskii & I. Gohberg, *The Decomposition of Figures into Smaller Parts*,

V. G. Boltyanskii & I. Gohberg, *Results and Problems in Combinatorial Geometry*, Cambridge University Press, Cambridge, 1985; *MR* **87c**:52002.

B. Grünbaum, Borsuk's problem and related questions, in [Kle], 271–284; *MR* **27** #4134.

H. Hadwiger, Ungelöste Probleme Nr. 20, *Elem. Math.* **12** (1957) 121.

S. Krotoszyński, Covering a plane convex body with five smaller homothetical copies, *Beiträge Algebra Geom.* **25** (1987) 171–176; *MR* **88g**:52011.

M. Lassak, Solution of Hadwiger's covering problem for centrally symmetric convex bodies in E^3, *J. London Math. Soc.* (2) **30** (1984) 501–511; *MR* **87e**:52024.

M. Lassak, Covering a plane convex body by four homothetical copies with the smallest positive ratio, *Geom. Dedicata* **21** (1986) 157–167; *MR* **88c**:52013.

M. Lassak, Covering the boundary of a convex set by tiles, *Proc. Amer. Math. Soc.* **104** (1988) 269–272; *MR* **89f**:52038.

M. Lassak, Covering plane convex bodies with smaller homothetical copies, in [BF], 331–337; *MR* **88i**:52023.

F. W. Levi, Überdeckung eines Eibereiches durch Parallelverschiebungen seines offenen Kerns, *Arch. Mat.* **6** (1955) 369–370; *MR* **17**, 888.

W. Moser & J. Pach, Problem 31, [MP].

C. A. Rogers, [Rog].

O. Schramm, Illuminating sets of constant width, *Mathematika* **35** (1988) 180–189.

P. S. Soltan, Covering of convex bodies by large homothetic ones, *Mat. Zametki.* **12** (1972) 85–90; *MR* **47** #7603.

D18. **The worm problem.** Leo Moser asked what are the minimal comfortable living quarters for a "unit worm"? More precisely, it is required to find the convex set K of least area that contains a congruent copy of every continuous (rectifiable) curve of length 1. There are many possible formulations of this problem. We can summarize the problems by three "parameters," the first being the kind of equivalent copy that we are interested in (e.g., congruence or translation), the second being any restrictions on the set K, and the third any condition on the class of unit worms considered.

For the most basic problem [congruence, all convex sets, all unit worms] the smallest cover yet discovered appears to be a certain truncated rhombus of area less than 0.286... (Figure D7) due to Gerriets and Poole. The problem [translation, all convex sets, all unit worms] was solved many years ago by Pál—the mimimal set is the equilateral triangle of height 1. The case of [translation, all convex sets, closed worms] was studied by Wetzel, who showed that the minimum area must be at least 0.155... and recently Bezdek & Connelly found a cover of area 0.165.... (A **closed worm** is one with its tail in its mouth.) Besicovitch studied the problems [congruence, equilateral

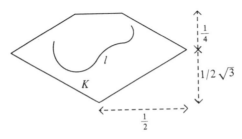

Figure D7. The truncated rhombus that provides the minimal cover for all curves of length one.

triangle, all worms] and [congruence, equilateral triangle, convex arcs], and, in two papers, Wetzel discussed [translation/congruence, any triangle, closed worms], where the minimal triangles are equilateral, and [congruence, circular sections, all worms/closed worms].

We can also seek the convex sets of minimal perimeter length or of minimal diameter in each case (see Bezdek & Connelly for a table of known results).

We can ask the opposite type of question, for example, what is the largest number $L(a, b, c)$ such that every curve of length $L(a, b, c)$ may be accommodated in a triangle of sides a, b, and c; how does L vary with a, b, and c? Clearly we can concoct many variations on this—the true mathematician will of course seek problems that have elegant or ingenious solutions!

There are obvious 3-dimensional analogs of these problems, but very little seems to be known about the least volume of covering sets of certain types.

This problem is considered for non-convex sets in Section G6.

A. S. Besicovitch, On arcs that cannot be covered by an open equilateral triangle of side 1, *Math. Gaz.* **49** (1965) 286–288; *MR* **32** #6320.

K. Bezdek & R. Connelly, Covering curves by translates of a convex set, *Amer. Math. Monthly* **96** (1989) 789–806.

G. D. Chakerian & M. S. Klamkin, Minimal covers for closed curves, *Math. Mag.* **46** (1973).

J. Gerriets & G. Poole, Minimum covers for arcs of constant length, *Bull. Amer. Math. Soc.* **79** (1973) 462–463; *MR* **47** #4150.

J. Gerriets & G. Poole, Convex regions which cover arcs of constant length, *Amer. Math. Monthly* **81** (1974) 36–41; *MR* **48** #12310.

J. C. C. Nitsche, The smallest sphere containing a rectifiable curve, *Amer. Math. Monthly* **78** (1971) 881–882; *MR* **45** #480.

J. Pál, Ein minimumprobleme für ovale, *Math. Ann.* **83** (1921) 311–319.

J. Schaer & J. E. Wetzel, Boxes for curves of constant length, *Israel J. Math.* **12** (1972) 257–265; *MR* **47** #5726.

J. E. Wetzel, Triangle covers for closed curves of constant length, *Elem. Math.* **25** (1970) 78–82; *MR* **42** #960.

J. E. Wetzel, On Moser's problem of accommodating closed curves in triangles, *Elem. Math.* **27** (1972) 35–36; *MR* **45** #4282.

J. E. Wetzel, Sectorial covers for curves of constant length, *Canad. Math. Bull.* **16** (1973) 367–375; *MR* **50** #14451.

E. Combinatorial Geometry

In this chapter we are concerned with the relative positions of several (usually convex) sets in the plane or in space. The first few sections are related to Helly's theorem, a result concerning the possible intersections among a collection of convex sets. This chapter also includes a number of problems on the placing of sets in the plane relative to the square lattice.

L. Danzer, B. Grünbaum & V. Klee, [DGK].
H. Hadwiger, H. Debrunner & V. Klee, [HDK].

E1. Helly-type problems. Helly's theorem states that if $\{K_i\}$ is a finite family of convex sets in $\mathbb{R}^d$ with the property that any $d + 1$ have a common point, then there is a point common to them all. Helly's theorem has given rise to a vast number of variants and generalizations, known as Helly-type theorems, many of which are discussed in the survey by Danzer, Grünbaum & Klee and the book by Hadwiger, Debrunner & Klee. We give a few of the many unsolved questions below.

(a) Bárány, Katchalski & Pach prove the quantitative result that there is a constant $\alpha_d > 0$ such that if every $2d$ of the $\{K_i\}$ have a common volume of at least 1 then the volume common to the entire collection is at least α_d. They point out that the number $2d$ cannot be reduced, but they conjecture that their value of $\alpha_d = d^{-2d^2}$ can be reduced to d^{-cd} for some constant c. Similarly, they show that if the intersection of every $2d$ of a collection of convex sets has a diameter of at least 1 then the intersection of all the sets has diameter at least $\beta_d = d^{-2d}\sqrt{(d + 1)/2d}$, and conjecture that $\beta_d = cd^{-1/2}$ is the correct order of magnitude.

(b) Sallee and also Buchman & Valentine showed that if every $d + 1$ of a family of compact convex sets in $\mathbb{R}^d$ contain a common set of width w, then

131

there is a set of width w that is contained in all members of the family. They ask what additional conditions are required to ensure that the sets all contain some body of constant width w. (Of course a convex set of width w need not contain a subset of constant width w.) They also ask about intersections with sets of constant width: if every $d + 1$ members of a family of convex sets in $\mathbb{R}^d$ intersect some set of constant width w, what extra conditions are required to guarantee that there is a set of constant width w that intersects them all?

(c) We define the **piercing number** or **Gallai number** $p(k)$ of a class of convex sets in $\mathbb{R}^d$ to be the least number with the following property: Given a family of such sets with every k member having a common point, then the sets may be divided into $p(k)$ subfamilies, each with nonempty intersection. Intuitively, if every k of the sets can be pierced by a single pin, then the entire collection can be anchored by $p(k)$ pins. Helly's theorem is just that $p(d + 1) = 1$ for any collection of convex sets in $\mathbb{R}^d$.

If the family consists of translates of some convex body K, then the problem is equivalent to a certain rather technical covering problem (see Section 7 of Danzer, Grünbaum & Klee). For this situation in the plane, Grünbaum has shown that $p(2) = 2$ if K is an affine regular hexagon, and that $p(2) \leq 3$ for any centro-symmetric set K in the plane. On the other hand, $p(2) \geq d + 1$ if K is a centro-symmetric strictly convex body in $\mathbb{R}^d$, so in particular $p(2) = 3$ if K is a strictly convex centro-symmetric plane set. Very little else is known even in the special case of translates. Does there exist a d-dimensional convex K with $p(2) > d + 1$ for families of translates of K? Find reasonable bounds for the values of $p(k)$ when $k \geq 3$ for a family of translates of a general convex body.

The first problem of this type appeared in Fejes Tóth's book: Given a collection of disks that overlap pairwise, what is the smallest number of pins needed to pierce the collection? Danzer eventually showed that four pins always sufficed, giving $p(2) = 4$ for a family of disks. Find, or estimate $p(k)$ for other families of homothetic copies of a convex set K. If an affinely regular hexagon can be inscribed in a plane convex K in such a way that K has parallel supporting lines at opposite vertices of the hexagon, then $p(2) \leq 7$ for any family of homothetic copies of K. However, in general only very poor bounds are known in the homothetic case, particularly in higher dimensions.

A much deeper discussion of these problems, with many further questions, is given in the article of Danzer, Grünbaum & Klee.

I. Bárány, M. Katchalski & J. Pach, Helly's theorem with volumes, *Amer. Math. Monthly* **91** (1984) 362–365; *MR* **86e**:52010.

I. Bárány, M. Katchalski & J. Pach, Quantitative Helly-type theorems, *Proc. Amer. Math. Soc.* **86** (1982) 109–114; *MR* **84h**:52016.

E. O. Buchman & F. A. Valentine, Any new Helly numbers? *Amer. Math. Monthly* **89** (1982) 370–375.

L. Danzer, Über Durchschnittseigenschaften n-dimensionalen Kugelfamilien, *J. Reine Angew. Math.* **208** (1961) 181–203; *MR* **25** #5453.

L. Danzer & B. Grünbaum, Intersection properties of boxes in R^d, *Combinatorica* **2** (1982) 237–246; *MR* **84g**:52014.

L. Danzer, B. Grünbaum & V. Klee, [DGK].

B. Grünbaum, Borsuk's partition conjecture in Minkowski planes, *Bull. Res. Council Israel Sect.* F **7** (1957) 25–30; *MR* **21** #2209.

B. Grünbaum, On intersections of similar sets, *Portugal. Math.* **18** (1959) 155–164; *MR* **23** #A2792.

H. Hadwiger & H. Debrunner, Über eine Variante zum Hellyschen Satz, *Arch. Math.* **8** (1957) 309–313; *MR* **19**, 1192.

H. Hadwiger, H. Debrunner & V. Klee, [HDK].

E. Helly, Über Mengen konvexer Körper mit gemeinschaftliches Punkten, *Jber. Deutsch. Math.-Verein* **32** (1923) 175–176.

G. T. Sallee, A Helly-type theorem for widths, in *The Geometry of Metric and Linear Spaces*, Lecture Notes in Math. 490, Springer, Berlin, 1975, 227–232; *MR* **53** #6425.

L. Stachó, A solution of Gallai's problem on pinning down circles, *Math. Lapok* **32** (1981) 19–47.

E2. **Variations on Krasnosel'skiĭ's theorem.** Suppose that a (2-dimensional) art gallery has the property that any three points are simultaneously observable by a suitably placed guard. Krasnosel'skiĭ's theorem guarantees that the guard may be positioned to observe the entire art gallery (see Figure E1).

Mathematically, let E be a subset of $\mathbb{R}^d$. We say that a point $\mathbf{x}$ of E *sees* a point $\mathbf{y}$ of E if the line segment $[\mathbf{x}, \mathbf{y}]$ lies entirely in E. If there is a point $\mathbf{x}$ that sees every point in E, we say that E is **starshaped**, and the set of $\mathbf{x}$ with this property is called the **kernel**, ker E, of E. The kernel is always a convex set. Krasnosel'skiĭ's theorem, an easy consequence of Helly's theorem (see Valentine) states that a compact set E in $\mathbb{R}^d$ is starshaped, if and only if for every set of $d + 1$ points of E, there is some point of E that sees all $d + 1$ points.

One can ask for other conditions involving relatively small sets of points that ensure starshapedness. It is easy to see that the conclusion of Krasnosel'skiĭ's theorem holds if every $d + 1$ boundary points of E be seen from a common point. The boundary points $\mathbf{x}$ for which $B_r(\mathbf{x}) \cap E$ is not convex for any $r > 0$ are called **locally non-convex** (LNC) points and are crucial to the "visibility structure" of E. It was shown by Breen that a compact

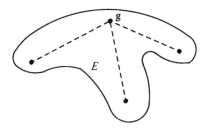

Figure E1. Krasnosel'skiĭ's theorem. If every three points of E are visible from a common point, then there is a point **g** in E that can see every point of E.

connected set E in $\mathbb{R}^2$ is starshaped if every three LNC points of E are clearly visible from a common point of E. (A point $\mathbf{x}$ is **clearly visible** from $\mathbf{y}$ if $\mathbf{y}$ sees all the points of $B_r(\mathbf{x}) \cap E$ for some ball $B_r(\mathbf{x})$ of center $\mathbf{x}$ and radius $r > 0$.)

A fundamental problem is whether such results hold in $\mathbb{R}^d$ for $d \geq 3$, that is whether in higher dimensions starshapedness is determined by the LNC points (see the survey by Kay).

Most of the work that has been done in this area relates to compact sets E. However, at least some results are possible if either the "closed" or the "bounded" conditions are relaxed. For example, any bounded set in $\mathbb{R}^2$ is starshaped if every three boundary points are clearly visible from a common point. What are the higher dimensional analogs?

A set E is **finitely starlike** if every finite set of points in E see a common point of E. (A compact set is finitely starlike, if and only if it is starshaped, but there are many non-compact sets that are finitely starlike without being starshaped, for example, $\{(x, y) \in \mathbb{R}^2 : y \leq \sin x\}$.) Peterson asks if there is an integer n such that if every n points of a finitely starlike E are visible from a common point, then E is starshaped. Breen has some partial results, and examples show that in the plane case $n \geq 9$, if it exists.

Many variations of Krasnosel'skii's theorem are concerned with properties of the kernel, ker E. In a series of papers, Breen has obtained conditions that ensure that the dimension of the kernel of a compact set E is not too small: dim ker $E \geq k$ if for some $\varepsilon > 0$, every $f(d, k)$ points of E see in E a common k-dimensional disk of radius ε, where $f(d, k) = 2d$ if $1 \leq k \leq d - 1$ and $f(d, d) = d + 1$. Can the stated values of $f(d, k)$ be reduced if $1 < k < d$? If $k = 1$ or $k = d$, this result is the best possible.

Motzkin (see Valentine, page 86) asks whether there is a Krasnosel'skii-type theorem for seeing objects of a finite size inside E. Let K be a (not too large) plane convex set. Suppose that, given any three points of a plane set E, we may translate K to a position inside E so that each of the three points sees some point of K. Can K be translated to a position so that every point of E sees some point of K?

Given $2 \leq r \leq d$, characterize the compact sets E in $\mathbb{R}^d$ such that every r points of E see a common point in E. If $r = 2$, E is called an L_2-**set** by Horn & Valentine, who show that if E is simply connected and $d = 2$ then E is a union of convex sets, each pair of which have a point in common.

M. Breen, The dimension of the kernel of a planar set, *Pacific J. Math.* **82** (1979) 15–21; *MR* **81h**:52006.

M. Breen, k-dimensional intersections of convex sets and convex kernels, *Discrete Math.* **36** (1981) 233–237; *MR* **84f**:52006.

M. Breen, Clear visibility, starshaped sets, and finitely starlike sets, *J. Geom.* **19** (1982) 183–196; *MR* **84g**:52013.

M. Breen, A Krasnosel'skii type theorem for nonclosed sets in the plane, *J. Geom.* **18** (1982) 28–42; *MR* **84a**:52007.

M. Breen, Clear visibility and the dimension of kernels of starshaped sets, *Proc. Amer. Math. Soc.* **85** (1982) 414–418; *MR* **83h**:52012.

M. Breen, A characterization theorem for bounded starshaped sets in the plane, *Proc. Amer. Math. Soc.* **94** (1985) 693–698; *MR* **86h**:52007.

M. Breen, A Krasnosel'skiĭ-type theorem for unions of two starshaped sets in the plane, *Pacific J. Math.* **120** (1985) 19–31; *MR* **87c**:52014.

M. Breen, Krasnosel'skiĭ-type theorems, [GLMP], 142–146; *MR* **87c**:52013.

M. Breen, Improved Krasnosel'skiĭ theorems for the dimension of the kernel of a starshaped set, *J. Geom.* **27** (1986) 174–179; *MR* **88c**:52005.

M. Breen, Some Krasnosel'skiĭ numbers for finitely starlike sets in the plane, *J. Geom.* **32** (1988) 1–12; *MR* **89i**:52010.

M. Breen, A weak Krasnosel'skiĭ theorem in $\mathbb{R}^d$, *Proc. Amer. Math. Soc.* **104** (1988) 558–562; *MR* **89i**:52011.

K. J. Falconer, The dimension of the kernel of a compact starshaped set, *Bull. London Math. Soc.* **9** (1977) 313–316; *MR* **57** #7392.

A. Horn & F. A. Valentine, Some properties of L-sets in the plane, *Duke Math. J.* **16** (1949) 131–140; *MR* **10**, 468.

D. C. Kay, Generalizations of Tietze's theorem on local convexity for sets in R^d, in [GLMP], 179–191; *MR* **87a**:52009.

M. A. Krasnosselsky, Sur un critère pour qu'un domaine soit étoilé, *Mat. Sbornik N.S.* **19 (61)** (1946) 309–310; *MR* **8**, 525.

B. B. Peterson, Is there a Krasnoselski theorem for finitely starshaped sets? [KB], 81–84.

N. Stavrakas, The dimension of the convex kernel and points of local nonconvexity, *Proc. Amer. Math. Soc.* **34** (1972) 222–224; *MR* **45** #7601.

F. A. Valentine, *Convex Sets*, McGraw-Hill, New York, 1964; *MR* **30** #503.

E3.   **Common transversals.** A straight line is called a **common transversal** of a family of sets in the plane if it cuts every set of the collection. We say that a collection of sets has property T if it has a common transversal, and property $T(n)$ if every subcollection of n subsets has a common transversal.

Vincensini raised the question of whether there is an integer n such that a family $\mathscr{F}$ of sets with property $T(n)$ necessarily has property T. Examples given by Hadwiger, Debrunner & Klee show that this is false for general families of convex sets: for each n there is a collection of plane convex sets such that every n have a common transversal, but such that no line cuts the entire collection. Lewis has given such an example with the sets as disjoint rectangles.

However, for restricted classes of sets, there can be an integer n such that $T(n)$ implies T. The least such n is called the **stabbing number** of the class (if any n sets can be stabbed by a line, then there is a line that skewers the lot!). Santaló showed that, for a collection of rectangles with sides parallel to the coordinate axes, if every six have a common transversal, then they all do. If the rectangles are also congruent and disjoint, Grünbaum showed that "six" can be replaced by "five." He also obtained a stabbing number of four for a collection of (at least six) disjoint circular disks of equal radii (see also Danzer). Grünbaum conjectures that a family of disjoint translates of any given convex set has a stabbing number of six. Katchalski shows that the stabbing number is at most 128; surely this number can be reduced. What can be said about other families of (congruent, translated, or homothetic) polygons? Can the stabbing number be bounded in terms of the number of sides?

There are many higher dimensional problems. A k-**transversal** for a family of sets is a k-dimensional plane in $\mathbb{R}^d$ that cuts all the sets. The k-**stabbing number** is the least integer n (if any) such that whenever every n-member subcollection of a family of convex sets of a given type has a k-transversal, then the entire collection has a k-transversal. [In the case of $k = 0$ the stabbing number of any collection of sets in $\mathbb{R}^d$ is $d + 1$; this is just Helly's theorem (see Section E1)]. Santaló showed that for a family of d-dimensional parallelepipeds with sides parallel to the coordinate axes, the 1-stabbing number is at most $(2d - 1)2^d$ and the $(d - 1)$-stabbing number is at most $2^{d-1}(d + 1)$, though the actual values are probably rather smaller. Little seems to be known about k-stabbing numbers for other k and for other classes of sets.

Kramer proved that if any three of a family of convex sets in the plane admit a common transversal, then there exist five lines with each member of the family intersected by at least one of them. Is the number five best possible? Eckhoff shows that at least three lines are required, and also examines this problem in the special case where the family consists of homothets of some convex set. Again, little seems to be known about higher-dimensional analogs.

L. Danzer, B. Grünbaum & V. Klee, [DGK].

L. Danzer, Über ein Problem aus der kombinatorischen Geometrie, *Arch. Math.* **8** (1957) 347–351; *MR* **19**, 977.

J. Eckhoff, Transversalenprobleme in der Ebene, *Arch. Math.* (*Basel*) **24** (1973) 195–202; *MR* **47** #9426.

B. Grünbaum, On common transversals, *Arch. Math.* **9** (1958) 465–469; *MR* **21** #1566.

B. Grünbaum, Common transversals for families of sets, *J. London Math. Soc.* **35** (1960) 408–416; *MR* **23** #A2793.

H. Hadwiger, H. Debrunner & V. Klee, [HDK].

M. Katchalski, Thin sets and common transversals, *J. Geom.* **14** (1980) 103–107; *MR* **82a**:52005.

M. Katchalski, A conjecture of Grünbaum on common transversals, *Math. Scand.* **59** (1986) 192–198; *MR* **88f**:52012.

T. Lewis, Two counterexamples concerning transversals for convex subsets of the plane, *Geom. Dedicata* **9** (1980) 461–465; *MR* **82e**:52006.

L. A. Santaló, Un teorema sôlve conjuntos de paralelepipedos de aristas paralelas, *Publ. Inst. Mat. Univ. Nac. Litoral* **2** (1940) 49–60; *MR* **2**, 261.

P. Vincensini, Figures convexes et variétés linéaires de l'espace euclidean à n dimensions, *Bull. Sci. Math.* **59** (1935) 163–174.

E4. Variations on Radon's theorem.

Radon's theorem, that any $d + 2$ points in $\mathbb{R}^d$ can be split into two subsets whose convex hulls have a common point, is fundamental to discrete combinatorial geometry. An important generalization was conjectured by Birch and proved by Tverberg, namely, that every set of $(m - 1)(d + 1) + 1$ points in $\mathbb{R}^d$ may be split into m disjoint subsets whose convex hulls have a common point. A good account of Radon's theorem and its equivalence to Helly's theorem is given by Danzer, Grünbaum & Klee, and more recent work is described in the surveys of Eckhoff and Reay.

One natural generalization of the theorems of Radon and Tverberg would provide information on the dimension of the intersection of the convex hulls. Reay conjectures that $2d(m - 1)$ points in $\mathbb{R}^d$ split into m subsets whose convex hulls contain a common line segment. He proves this when $d = 2$ and Eckhoff has given a proof for $m = 2$. At the other extreme, can $m(d + 1)$ points in general position (i.e., no three in a line, no four on a plane, etc.) be arranged as the vertices of m simplexes which have a common interior point? This was recently proved by Roudneff for $d = 3$, so in particular 16 points in general position in space can be partitioned into four tetrahedra with a common interior point. Is it true for $d = 4$, and $m = 3$? The most general conjecture of this type is that $(m - 1)(d + 1) + k + 1$ points in general position have a decomposition into m subsets with convex hulls intersecting in a convex set of dimension k; Roudneff has again proved this when $d = 3$.

In a similar vein, Doignon asks if every set of $2d + 1$ points in $\mathbb{R}^d$ can be split into two subsets S_1 and S_2 such that $\dim(\text{conv } S_1 \cap \text{conv } S_2) \geq \min\{\dim(\text{conv } S_1), \dim(\text{conv } S_2)\}$. Examples show that $2d + 1$ cannot be replaced by $2d$.

Birch asks whether every set of 8 points in the plane can be partitioned to form two triangles and a segment so that the segment cuts the interior of both triangles. More generally, if $m > t \geq 1$, can every set of $(m - 1)d + t + 1$ points in $\mathbb{R}^d$ be partitioned into m subsets in such a way that among the convex hulls of these sets there are t, each of which intersect all convex hulls? If $t = m - 1$ this is just Tverberg's theorem.

Many further problems and conjectures, some rather technical, may be found in the articles by Eckhoff and Reay (see also Reay's note in Moser & Pach, Problem 56).

L. Danzer, B. Grünbaum & V. Klee, [DGK].

J. P. Doignon, Quelques problèmes de convexité combinatoire, *Bull. Soc. Math. Belg.* Sér. B **31** (1979) 203–207; *M R* **83i**:52010.

J. Eckhoff, On a class of convex polytopes, *Israel J. Math.* **23** (1976) 332–336; *M R* **53** #14313.

J. Eckhoff, Radon's theorem revisited, in [TW], 164–185; *M R* **81f**:52016.

W. Moser & J. Pach, [MP], Problem 56.

B. B. Peterson, The geometry of Radon's theorem, *Amer. Math. Monthly* **79** (1972) 949–963; *M R* **47** #4143.

J. R. Reay, Several generalizations of Tverberg's theorem, *Israel J. Math.* **44** (1979) 238–244; *M R* **81f**:52017.

J. R. Reay, Twelve general-position points always form three intersecting tetrahedra, *Discrete Math.* **28** (1979) 193–199; *M R* **81c**:05021.

J. R. Reay, Open problems around Radon's theorem, in [KB], 151–172; *M R* **83f**:52006.

J.-P. Roudneff, Partitions of points into intersecting tetrahedra, *Discrete Math.* **81** (1990) 81–86.

H. Tverberg, A generalization of Radon's theorem, *J. London Math. Soc.* **41** (1966) 123–128; *M R* **32** #4601.

E5. Collections of disks with no three in a line.

Motzkin asked for good asymptotic estimates for the largest radius $r(n)$ of n equal disks that can be placed in the unit disk so that no straight line intersects more than two of them. It seems certain that some of the methods developed for Heilbronn's triangle problem (see Guy) are applicable here; in particular, the example of

Komlós, Pintz & Szemerédi shows that $r(n) \geq c \ln n/n^2$ compared with the obvious $r(n) \geq c/n^2$. Trivially $r(n) \leq c/n$, but is $r(n) \leq c/n^{2-\varepsilon}$ for all $\varepsilon > 0$?

More generally, how large can the radius of n equal disks in the unit disk be if no line cuts more than k of them?

There are various higher-dimensional analogs. In 3-dimensional space we can ask for the maximum radius of spheres contained in the unit sphere such that (a) no line intersects more than two, or (b) no plane intersects more than three.

R. K. Guy, *Unsolved Problems in Number Theory*, Problem F4, Springer, New York, 1981.

J. Komlós, J. Pintz & E. Szemerédi, A lower bound for Heilbronn's problem, *J. London Math Soc.* (2) **25** (1982) 13–24; *M R* 83i:10042.

E6. Moving disks around. Let $B_1, \ldots, B_n$ and $D_1, \ldots, D_n$ be disks in the plane of equal radii with centers $\mathbf{x}_1, \ldots, \mathbf{x}_n$ and $\mathbf{y}_1, \ldots, \mathbf{y}_n$, respectively. Suppose that the centers satisfy

$$|\mathbf{y}_i - \mathbf{y}_j| \leq |\mathbf{x}_i - \mathbf{x}_j| \qquad (1 \leq i < j \leq n). \tag{1}$$

Does it follow that

$$\text{area of } \bigcup_{i=1}^{n} D_i \leq \text{area of } \bigcup_{i=1}^{n} B_i \quad ? \tag{2}$$

This problem is due to Poulsen and Kneser, and has been discussed by Hadwiger, Valentine & Klee. Bollobás showed that Eq. (2) holds provided that the centers $\mathbf{x}_i$ can be moved continuously to $\mathbf{y}_i$ in such a way that the distances $|\mathbf{x}_i - \mathbf{x}_j|$ never increase. Unfortunately there are simple examples where the condition fails; however Bollobás's result includes the case treated by Bouligand where the sets $\{\mathbf{x}_1, \ldots, \mathbf{x}_n\}$ and $\{\mathbf{y}_1, \ldots, \mathbf{y}_n\}$ are similar. In the general case, Kneser shows that Eq. (2) becomes true if a factor of 9 is inserted on the right-hand side. Can one get a constant smaller than 9?

A "dual" problem is whether, under hypothesis (1), we have

$$\text{area of } \bigcap_{i=1}^{n} D_i \geq \text{area of } \bigcap_{i=1}^{n} B_i \quad ?$$

Many authors have shown that if $\bigcap_{i=1}^{n} B_i$ is nonempty then so is $\bigcap_{i=1}^{n} D_i$ (e.g., see Grünbaum or Alexander).

One can also consider these problems where the B_i and D_i are more general convex sets, but the situation becomes complicated (see Rehder).

Alexander has considered some further problems of this type. He shows that if Eq. (1) holds then the convex hull $\text{conv}\{\mathbf{x}_1, \ldots, \mathbf{x}_n\}$ has a perimeter length of at least that of $\text{conv}\{\mathbf{y}_1, \ldots, \mathbf{y}_n\}$ with a similar result for total mean curvature. Can we conclude that the area of $\text{conv}\{\mathbf{x}_1, \ldots, \mathbf{x}_n\}$ is at least the area of $\text{conv}\{\mathbf{y}_1, \ldots, \mathbf{y}_n\}$ if for all $1 \leq i \leq j \leq k \leq n$ we have $\text{area}(\Delta\mathbf{x}_i\mathbf{x}_j\mathbf{x}_k) \geq \text{area}(\Delta\mathbf{y}_i\mathbf{y}_j\mathbf{y}_k)$? [Trivial examples show that the area inequality for the convex hull does not follow from Eq. (1) alone.]

R. Alexander, The circumdisk and its relation to a theorem of Kirszbraun and Valentine, *Math. Mag.* **57** (1984) 165–169; *MR* **85f**:52028.

R. Alexander, Lipschitzian mappings and total mean curvature of polyhedral surfaces I, *Trans. Amer. Math. Soc.* **288** (1985) 661–678; *MR* **86c**:52004.

B. Bollobás, Area of the union of discs, *Elem. Math.* **23** (1968) 60–61; *MR* **38** #3772.

G. Bouligand, Ensembles impropres et nombre dimensional, *Bull. Sci. Math.* **52** (1928) 320–344.

L. Danzer, B. Grünbaum & V. Klee, [DGK].

M. Gromov, Monotonicity of the volumes of intersections of balls, in *Geometrical Aspects of Functional Analysis (1985/86)*, Lecture Notes in Math. 1267, Springer, Berlin, 1987, 1–4.

B. Grünbaum, A generalization of theorems of Kirszbraun and Minty, *Proc. Amer. Math. Soc.* **13** (1962) 812–814; *MR* **27** #6110.

H. Hadwiger, Ungelöste Probleme Nr. 11, *Elem. Math.* **11** (1956) 60–61.

V. Klee, Some unsolved problems in plane geometry, *Math. Mag.* **52** (1979) 131–145; *MR* **80m**:52006.

M. Kneser, Eine Bemerkung über des Minkowskische Flächenmass, *Arch. Math.* **6** (1955) 382–390; *MR* **17**, 469.

G. J. Minty, A finite-dimensional tool-theorem in monotone operator theory, *Adv. Math.* **12** (1974) 1–7; *MR* **49** #5952.

W. Moser & J. Pach, Problem 39, [MP].

E. T. Poulsen, Problem 10, *Math. Scand.* **2** (1954) 376.

W. Rehder, On the volume of unions of translates of a convex set, *Amer. Math. Monthly* **87** (1980) 382–384; *MR* **81g**:52004.

F. A. Valentine, *Convex Sets*, McGraw-Hill, New York, 1964; *MR* **30** #503.

E7. **Neighborly convex bodies.** An arrangement of convex polyhedra is called **neighborly** if every pair of the polyhedra abut in a region of positive area. How many tetrahedra can there be in a neighborly collection? Bagemihl gave an example of eight such tetrahedra, and conjectured that this was the greatest possible number. Baston showed that no ten tetrahedra could be neighborly. Recently, Zaks showed that nine was also impossible, thus proving the conjecture. Zaks's proof involves showing that a set of diophantine relations has no solution, and involves a substantial amount of computation. A shorter demonstration would be very welcome. Zaks has also found arrangements of 2^d neighborly simplexes in $\mathbb{R}^d$ for $d \geq 4$ (in this case we require that every pair of simplexes have a $(d - 1)$-dimensional region in common). It is widely believed that no arrangement of more than 2^d simplexes can be neighborly, generalizing the 3-dimensional situation. However, the best result known is due to Perles, who showed that any neighborly family of d-simplexes contains at most 2^{d+1} members, and more generally, that a neighborly family of d-polytopes each with at most k faces, contains at most 2^k members.

It is easy to see that it is impossible to find five regions in the plane (each bounded by a rectifiable curve) such that each pair has a positive length of boundary in common. On the other hand, four such regions certainly exist, and can be convex. In $\mathbb{R}^3$ this problem becomes more interesting. Tietze answered a question of Stächel by constructing an infinite family of convex bodies in $\mathbb{R}^3$ with each pair having boundary contact of positive area. Crum posed the problem independently some forty years later, and Besicovitch gave a second construction. This problem has become known as "Crum's problem" and is discussed by Tietze and by Danzer, Grünbaum & Klee. This last article describes a neat dual construction from four polytopes that gives a neighborly family of polyhedra with the polyhedra touching in common faces. Dewdney & Vranch have given an elegant argument to show that the Voronoi regions of the positive integer points (n, n^2, n^3) on the moment curve provide a solution to Crum's problem. (The **Voronoi region** of x is that part of space nearer to x than to any of the other given points.) They ask whether points on any "twisted" curve (i.e., with no four coplanar points) can be used in the construction. Zaks has used related methods to answer a question of Grünbaum, by showing that if $d \geq 3$, there exist arbitrarily large families of *centro-symmetric* convex polytopes in $\mathbb{R}^d$. Indeed, if $d \geq 4$, they can all be congruent. Some other generalizations of Crum's problem are discussed by Rado and Eggleston.

Similar questions on the size of neighborly families may be asked when other conditions are imposed. For example, it is known that if the neighborly polyhedra are all combinatorially equivalent to a cube, then the maximum is between 24 and 64; if they are all prisms, then the bounds are 16 and 32. What are the exact maximum values?

A question of particular interest is what is the maximum number of *congruent* polyhedra that may be arranged in a neighborly fashion. Presumably, the maximum is finite, though this has not been proved. Zaks has given an elegant example of a neighborly family of eight congruent triangular prisms: take the prism P with vertices (0, 0, 0), (1, 0, 0), (0, 1, 0), (0, 0, 5), (1, 0, 5,), and (0, 1, 6) (note that the top of P makes an angle $\pi/4$ with the base). Place around P three congruent copies of P each with one face on the (x, y)-plane and lying above this plane. Then these four polyhedra are neighborly. It is then possible to position an identical arrangement of prisms on the other side of the (x, y)-plane so that each of these has an area in common with each of the first four. Can one arrange nine, or even more, congruent polyhedra in a neighborly fashion?

How many solid congruent convex bodies in $\mathbb{R}^3$ can be placed so that each pair has some *point* of contact? (To avoid trivial examples, we require each pair to touch at a different point). Presumably, the maximum number is finite, but the above example shows that it is at least eight. A configuration of seven mutually touching congruent circular cylinders has been known for some time (a good use for cigarettes!), and Littlewood asked whether seven *infinite* cylinders of unit radius can be arranged to touch each other. This is equivalent to the problem of whether seven lines can be drawn in space at unit (perpendicular) distance from each other.

Zaks defines a family of polyhedra to be **nearly-neighborly** if, for every two members, there exists a plane that separates them and contains a face of each. There can be at most 14 nearly-neighborly tetrahedra in a collection, but Zaks conjectures that 8 is the actual maximum.

Several other "neighborly" problems are discussed in the American Mathematical Monthly article by Zaks.

F. Bagemihl, A conjecture concerning neighboring tetrahedra *Amer. Math Monthly* **63** (1956) 328–329; *MR* **17**, 995.

V. J. D. Baston, *Some Properties of Polyhedra in Euclidean Space*, Pergamon, Oxford, 1965; *MR* **31** #5127.

A. S. Besicovitch, On Crum's problem, *J. London Math. Soc.* **22** (1947) 285–287; *MR* **9**, 605.

L. Danzer, B. Grünbaum & V. Klee, [DGK], in particular Section 8.

A. K. Dewdney & J. K. Vranch, A convex partition of R^3 with applications to Crum's problem and Knuth's post-office problem, *Utilitas Math.* **12** (1972) 193–199; *MR* **58** #2605.

H. G. Eggleston, On Rado's extension of Crum's problem, *J. London Math. Soc.* **28** (1953) 467–471; *MR* **15**, 248.

M. Gardner, *Mathematical Puzzles and Diversions*, Simon and Schuster, New York, 1958, 115.

C. J. A. Halberg, E. Levin & E. G. Straus, On contiguous congruent sets in Euclidean space, *Proc. Amer. Math. Soc.* **10** (1959) 335–344; *MR* **21** #5170.

V. Klee, Can nine tetrahedra form a neighboring family? *Amer. Math. Monthly* **76** (1969) 178–179.

J. E. Littlewood, *Some Problems in Real and Complex Analysis*, Heath, Boston, 1968, 20; *MR* **39** #5777.

W. Moser & J. Pach, Problem 55, [MP].

M. A. Perles, At most 2^{d+1} neighborly simplices in E^d, in [RZ], 253–254; *MR* **87b**:52026.

R. Rado, A sequence of polyhedra having intersections of specified dimensions, *J. London Math. Soc.* **22** (1947) 287–289; *MR* **9**, 605.

H. Tietze, Über Nachbargebiete in Raume, *Monatsch. Math.* **16** (1905) 211–216.

H. Tietze, *Gelöste und Ungelöste Mathematische Probleme aus Alter und Neuer Zeit*, Biederstein, Munich, 1949; *MR* **11**, 571.

J. Zaks, Neighborly families of 2^d d-simplices in E^d, *Geom. Dedicata* **11** (1981) 505–507; *MR* **83a**:52011.

J. Zaks, A solution to Bagemihl's conjecture, *C. R. Math. Rep. Acad. Sci. Canada* **8** (1986) 317–321.

J. Zaks, No nine neighborly tetrahedra exist, to appear.

J. Zaks, Arbitrarily large neighborly families of symmetric convex polytopes, *Geom. Dedicata* **20** (1986) 175–179; *MR* **87f**:52013.

J. Zaks, Neighborly families of congruent convex polytopes, *Amer. Math. Monthly* **94** (1987) 151–155; *MR* **88b**:52018.

J. Zaks, Nearly-neighborly families of tetrahedra and the decomposition of some multigraphs, *J. Combin. Theory Ser. A* **48** (1988) 147–155.

E8. Separating objects.

A popular game, which has been marketed under various names, requires players to remove objects from a pile without disturbing the remaining objects. This suggests problems of the following nature (see de Bruijn). Given a finite configuration of non-overlapping bodies in $\mathbb{R}^d$, when can we move at least one of the bodies some short distance with the others remaining fixed? Similarly, when can one of them be removed to an arbitrary distance away?

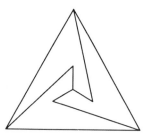

Figure E2.   Three plane sets, none of which can be removed without disturbing the others but which can be separated by simultaneous motion.

Fejes Tóth & Heppes exhibited an arrangement of 14 convex bodies in $\mathbb{R}^3$ none of which can be moved independently of the rest and Danzer and also Dawson gave examples with twelve congruent convex bodies. Is there a "self-blocking" configuration in $\mathbb{R}^3$ with fewer than twelve convex sets? Do configurations exist with all the sets smooth? Dawson showed that at least $\min(n, d + 1)$ spheres may be removed from an arrangement of n spheres in $\mathbb{R}^d$ without disturbing any of the others and also that any configuration of at least three convex sets in the plane contains three removable sets.

What happens if we are only allowed to translate the objects (without rotation)? Toussaint showed that given a finite set of spheres and a fixed direction, the spheres may all be translated a unit distance in this direction, one at a time, in some order without interpenetration. Post has an example of six convex sets in $\mathbb{R}^3$, none of which can be translated in any direction without collision. Is there an example with just five sets? Dawson writes that this is equivalent to the $d = 3$, $n = 5$ case of the "tangled simplex" problem: Do there exist n simplexes in $\mathbb{R}^d$ such that each face of each simplex has face-to-face contact with some face of another simplex? (Compare the neighborly tetrahedra problem in Section E7.) The answer is "no" in $\mathbb{R}^2$ for all n.

Dropping the convexity condition, Fejes Tóth & Heppes give neat configurations of three plane sets, none of which is removable (see Figure E2). Dawson has made a start on a "local analysis" of the general problem of removability and his paper is illustrated with nice examples of the ways in which objects can "interlock" if their boundaries are irregularly shaped.

For related problems and further references, see Section G5.

N. G. de Bruijn, Problems 17 and 18, *Nieuw Arch. Wisk.* (3) **2** (1954) 67.

L. Danzer, To Problem 8, 312 [Fen].

R. J. McG. Dawson, On removing a ball without disturbing the others, *Math. Mag.* **57** (1984) 27–30; *MR* **86b**:52005.

R. J. McG. Dawson, On the mobility of bodies in R^n, *Math. Proc. Cambridge Philos. Soc.* **98** (1985) 403–412, corrigenda **99** (1986) 377–379; *MR* **87h**:52016a,b.

L. Fejes Tóth & A. Heppes, Über stabile Körpersysteme, *Composito Math.* **15** (1963) 119–126; *MR* **28** #4435.

L. Fejes Tóth, Problem 8, 312 [Fen].

K. A. Post, Six interlocking cylinders with respect to all directions, Internal report, Eindhoven, 1983.

G. T. Toussaint, On translating a set of spheres, Tech. Rep. 84.4, McGill University, Montreal, 1984.

E9. Lattice point problems.

The next few problems involve properties of sets in relation to lattice points. For our purposes a lattice point will always be taken to mean a point in $\mathbb{R}^d$ with integer Cartesian coordinates. The **square lattice** or **integer lattice** will refer to the set of all such lattice points in $\mathbb{R}^d$ [see Figure E3(a)].

It is possible to pose lattice point problems for lattices other than the square one [see Figure E3(b)]. Sometimes such problems can be shown to be equivalent to that for the square lattice, though often different problems may result.

A large collection of unsolved problems on lattice points may be found in Hammer's book.

J. Hammer [Ham].

E10. Sets covering constant numbers of lattice points.

Steinhaus has asked whether there exists a plane set E that covers exactly n lattice points however it is placed on the lattice (rotations being allowed). One might hope to construct such an E by using the axiom of choice or other appropriate axioms of logic, but apparently no one has yet succeeded in doing this. (This is a particular case of the problem in Section G9.)

If we require E to be a Borel or measurable set, the problem takes on a completely different character. Certainly the area of such an E must equal n. The problem is equivalent to seeking a set such that the 2-dimensional Fourier transform of the characteristic function of E is zero on all circles centered on the origin with radii $\sqrt{i^2 + j^2}$, where i and j are positive integers. Croft has used the idea of the density boundary to show that E cannot be

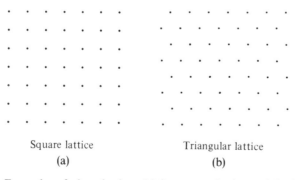

Square lattice   Triangular lattice
(a)   (b)

Figure E3. Examples of plane lattices: (a) for square lattice and (b) for triangular lattice. Our problems generally concern the square lattice, consisting of points with integer coordinates.

measurable and bounded, but his argument does not extend to the unbounded case.

Croft also poses the following problem. Put a measure on the set of congruence transformations ρ using some 3-dimensional parameterization (x, y, θ) in the natural way. Let E be plane-measurable. We say that the integer k is **essentially represented** by E if the set of ρ for which $\rho(E)$ contains k points has positive measure (i.e., if such covering positions are non-exceptional). Then is the set of integers essentially represented by E always a block of consecutive integers? This is not true if "essentially" is omitted: the closed disk of radius $\sqrt{5}$ can be moved to cover 17 or 21 lattice points, but not 18, 19, or 20.

H. T. Croft, Three lattice point problems of Steinhaus, *Quart. J. Math. Oxford Ser.* **33** (1982) 71–83; *MR* **85g**:11051.

I. Niven & H. S. Zuckerman, Lattice points in regions, *Proc. Amer. Math. Soc.* **18** (1967) 364–370; *MR* **35** #136.

H. Steinhaus, *Mathematical Snapshots*, Oxford University Press, Oxford, 1969.

E11. **Sets that can be moved to cover several lattice points.** It is well-known that every plane (measurable) set of area greater that $\frac{1}{4}\pi$ may be moved (rigidly) to a position so that it covers at least two points of the unit square lattice. What is the critical area for covering at least three, and more generally, at least n points?

A simple argument shows that any set E of infinite area (i.e., infinite inner measure) can be positioned to cover at least n lattice points for each integer n. A question sent to us by Steinhaus asks whether E can always be positioned to cover infinitely many lattice points. If not, what additional hypotheses are required for this to be so? Croft has shown that it is enough for E to have unbounded interior.

One can ask similar questions for convex sets. The following problem from L. Moser's collection is of particular interest. Let $f(A)$ be the largest number such that every plane convex set of area A can be positioned to cover $A + f(A)$ lattice points. Does $f(A) \to \infty$ as $A \to \infty$ and if so, how fast? Recently, Beck has shown that there is always a position covering more than $A + cA^{1/9}$ points and also one covering fewer than $A - cA^{1/9}$. He also points out that, using his method, the exponent $\frac{1}{9}$ can be replaced by $\frac{1}{8} - \varepsilon$ for any $\varepsilon > 0$. Simple examples show that $f(A) = O(A^{1/2})$, and Beck conjectures that $f(A) = O(A^{1/4})$ is the right order or magnitude.

We propose the following strengthening of Moser's problem: Can a convex set of area A be placed so as to intersect the parallel line set $\{(x, y): x$ is an integer$\}$ in total length at least $A + f(A)$ for some $f(A)$ which tends to infinity with A?

Of course, by placing further restrictions on the class of sets (e.g., circles, triangles, squares, n-gons, etc.,) we get a plethora of further problems (see Moser & Pach, Problem 53 for a discussion of the case of a square, and, for example, Kendall & Rankin in the case of a disk or sphere).

J. Beck, Irregularities of distribution, in *Surveys in Combinatorics*, London Math. Soc. Lecture Notes 103 (1985) 25–46; *MR* **87d**:05058.

J. Beck, On a lattice point problem of L. Moser I, II, *Combinatorica* **8** (1988) 21–47, 159–176.

H. T. Croft, Three lattice point problems of Steinhaus, *Quart. J. Math. Oxford Ser.* (2) **33** (1982) 71–83; *MR* **85g**:11051.

D. G. Kendall & R. A. Rankin, On the number of points of a given lattice in a random hypersphere, *Quart. J. Math. Oxford Ser.* (2) **4** (1953) 178–189; *MR* **15**, 237.

W. Moser & J. Pach, Problems 52 and 53, [MP].

R. Schark & J. M. Wills, Translationen, Bewegungen und Gitterpunktanzahl konvexer Bereiche, *Geom. Dedicata* **3** (1974) 251–256; *MR* **50** #4498.

E12. Sets that always cover several lattice points. Santaló asked for the convex set of least area with the property that every congruent copy contains at least one lattice point. Schäffer and Sawyer showed independently that the smallest such set is $K = \{(x, y): |x| \le \frac{1}{2}, |y| \le \frac{3}{4} - x^2\}$ which has area $\frac{4}{3}$. According to Reich, Schäffer conjectures that K is also the covering set of least perimeter. What is the (measurable) set of minimal area if the convexity condition is dropped—is it still the same set or can area $1 + \varepsilon$ be achieved for any $\varepsilon > 0$?

Describe the sets, convex or otherwise, of least area or of least perimeter with all congruent copies covering at least n lattice points for $n = 2, 3, \ldots$. Presumably for large n they approach a circular shape—are there any n for which they are circular? Mögling gives exact conditions for regular p-gons always to cover a lattice point if p is odd. Characterize such regular polygons for even p, and more generally find conditions for arbitrary p-gons always to cover n lattice points.

It is also useful to have conditions involving other measures of a convex set K that ensure that K contains at least one (more generally, at least n) lattice points however it is situated. The intuitive idea here is that if K can cover few points compared with its area A, then it must be "long and thin." A basic result in this direction is due to Nosarzewska who showed that however K is placed it must cover at least $A - \frac{1}{2}L$ and at most $A + \frac{1}{2}L + 1$ lattice points, where L is the perimeter of K. (See the papers of Bokowski, Hadwiger & Wills, of Schmidt and of Wills for higher dimensional analogs.)

Bender proved the "isoperimetric" result that if $\frac{1}{2}L \le A$ then K must contain a lattice point, and he extended this to higher dimensions. How big must A/L be to guarantee catching n lattice points? Scott showed that if $(w - 1)(D - 1) \ge 1$ then K contains some lattice point, and if $(w - \sqrt{2})(D - \sqrt{2}) \ge 2$ then K contains at least two points. (See McMullen & Wills for higher-dimensional versions of this.) What are the analogs for n points? Scott suggests that the more natural thing to look at here rather than w and D is what he terms the axial diameter. Scott also showed that a convex set satisfying $A > 1.144nD$ must contain n lattice points, and by a simple refinement of his argument, Hammer pointed out that such a set in fact contains n^2 points! Can these results be improved, or generalized to d dimensions for $d \ge 3$?

E. A. Bender, Area–perimeter relations for two dimensional lattices, *Amer. Math. Monthly* **69** (1962) 742–744.

K. Bezdek, The thinnest holding-lattice of a set, *Monatsh. Math.* **103** (1987) 177–185; *MT* **88j**:52031.

J. Bokowski, H. Hadwiger & J. M. Wills, Eine Ungleichung zwischen Volumen, Oberfläche und Gitterpunktzhal konvexer Körper in n-dimensionalen euklidischen Raum, *Math. Z.* **127** (1972) 363–364; *MR* **47** #4144.

H. T. Croft, Three lattice point problems of Steinhaus, *Quart. J. Math. Oxford Ser.* (2) **33** (1982) 71–83; *MR* **85g**:11051.

H. Hadwiger, Bemerkungen über Gitter und Volumen, *Mathematica Timişoara* **18** (1942) 97–103; *MR* **4**, 112.

J. Hammer, Lattice points and area-diameter relation, *Math. Mag.* **52** (1979) 25–26; *MR* **80d**:10044.

J. Hammer, [Ham].

P. McMullen & J. M. Wills, Minimal width and diameter of lattice-point-free convex bodies, *Mathematika* **28** (1981) 255–264; *MR* **83d**:52011.

E. A. Maier, On the minimal rectangular region which has the lattice point covering property, *Math. Mag.* **42** (1969) 84–85; *MR* **39** #6178a.

I. Niven & H. S. Zuckerman, Lattice point coverings by plane figures, *Amer. Math. Monthly* **74** (1967) 353–362; 952; *MR* **35** #1557, **36** #2557.

I. Niven & H. S. Zuckerman, The lattice point covering theorem for rectangles, *Math. Mag.* **42** (1969) 85–86; *MR* **39** #6178b.

M. Nosarzewska, Évaluation de la différence entre l'aire d'une région plane convexe et le nombre des points avec coordonnées entières couvert par elle, *Colloq. Math.* **1** (1948) 305–311; *MR* **11**, 13.

S. Reich, Comment on Problem 780, *Math. Mag.* **46** (1973) 51–52.

L. A. Santaló, Geometria integral de figuro ilimitadas, *Publ. Inst. Mato. Rosario* **1** (1942) 54.

D. B. Sawyer, On the covering of lattice points by convex regions II, *Quart. J. Math.* (2) **6** (1955) 207–212.

J. J. Schäffer, Smallest lattice-point covering set, *Math. Ann.* **129** (1955) 265–273; *MR* **16**, 1145.

W. M. Schmidt, Volume, surface area and the number of integer points covered by a convex set, *Arch. Math.* (*Basel*) **23** (1972) 537–543; *MR* **47** #5730.

P. R. Scott, Area-diameter relations for two-dimensional lattices, *Math. Mag.* **47** (1974) 218–221; *MR* **50** #1130.

P. R. Scott, Two inequalities for convex sets with lattice point constraints in the plane, *Bull. London Math. Soc.* **11** (1979) 273–278; *MR* **81a**:52012.

P. R. Scott, Area, width and diameter of plane convex sets with lattice point constraints, *Indian J. Pure Appl. Math.* **14** (1983) 444–448: *MR* **84j**:52015.

P. R. Scott, On planar convex sets containing one lattice point, *Quart. J. Math. Oxford Ser.* (2) **36** (1985) 105–111; *MR* **86d**:52007.

P. R. Scott, Lattices and convex sets in space, *Quart. J. Math. Oxford Ser.* (2) **36** (1985) 359–362; *MR* **86j**:52018.

P. R. Scott, On the area of planar sets containing many lattice points, *Bull. Austral. Math. Soc.* **35** (1987) 441–454; *MR* **88e**:52017.

J. M. Wills, Zur Gitterpunktenzahl Konvexer Mengen, *Elem. Math.* **28** (1973) 57–67; *MR* **53** #3898.

E13. Variations on Minkowski's theorem.

Let K be a convex set in $\mathbb{R}^d$ that is centro-symmetric about the origin **o**. If K has d-dimensional volume $V \geq 2^d$, then K contains at least two lattice points other than **o** (see Figure E4). This is Minkowski's theorem, which is of fundamental importance in the geometry

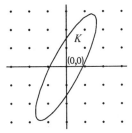

Figure E4.   Minkowski's theorem: If K is centro-symmetric about the origin and has an area of at least 4, it contains further lattice points.

of numbers (see Hardy & Wright or Cassels for proofs and consequences of this result). Many problems on convex bodies and lattice points come from trying to generalize Minkowski's theorem or varying the hypotheses. We mention some of these here, others may be found in Hammer's book and in the various papers of Scott.

Ehrhart conjectures that if K is any (not necessarily symmetric) convex body in $\mathbb{R}^d$ with centroid at $\mathbf{o}$ and containing no further lattice points, then $V \leq (d + 1)^d/d!$, this bound being attained by a certain simplex. He proved this in the plane case. Must sets of any larger area or volume always contain a pair of lattice points symmetric with respect to $\mathbf{o}$? Might it also be true that if $V \geq k(d + 1)^d/d!$ then K contains $2k$ lattice points situated symmetrically about $\mathbf{o}$?

The extremal convex set for Ehrhart's result in the plane is the triangle with vertices $(-2, -1)$, $(2, 1)$, and $(1, -1)$. Scott asks if this is also the convex set of greatest width with no interior lattice points other than $\mathbf{o}$.

Scott also asks for the convex set of largest area with circumcenter at $\mathbf{o}$ containing no other interior lattice points. He conjectures that the maximum is $4.04\ldots$ with the intersection of the circle center $\mathbf{o}$ and radius $1.637\ldots$ and a triangle with two lattice points on each side as the extremal.

Scott proposed maximizing A/L for a plane centro-symmetric convex set with center $\mathbf{o}$ and containing no other lattice points. This problem was solved by Arkinstall & Scott, and by Croft who considered the more general case of A/L^α for $0 < \alpha < 2$. The 3-dimensional analogs promise to be awkward!

J. R. Arkinstall, Minimal requirements for Minkowski's theorem in the plane I, II, *Bull. Austral. Math. Soc.* **22** (1980) 259–274, 275–283; *MR* **82f**:52013.

J. R. Arkinstall & P. R. Scott, An isoperimetric problem with lattice point constraints, *J. Austral. Math. Soc. Ser. A* **27** (1979) 27–36; *MR* **80h**:52011.

J. W. S. Cassels, *Introduction to the Geometry of Numbers*, Springer-Verlag, New York, 1971; *MR* **46** #5257.

H. T. Croft, Cushions, cigars and diamonds: an area-perimeter problem for symmetric ovals, *Math. Proc. Cambridge Philos. Soc.* **85** (1979) 1–16; *MR* **80e**:52001.

E. Ehrhart, Une généralization du théorème de Minkowski, *C. R. Acad. Sci. Paris* **240** (1955) 483–485; *MR* **16**, 574.

E. Ehrhart, Sur les ovales et les ovoides, *C. R. Acad. Sci. Paris* **240** (1955) 583–585; *MR* **16**, 740.

E. Ehrhart, Une généralization probable du théorème fondamental de Minkowski, *C. R. Acad. Sci. Paris* **258** (1964) 4885–4887; *MR* **29** #522.

J. Hammer, Some relatives of Minkowski's theorem for two-dimensional lattices, *Amer. Math. Monthly* **73** (1966) 744–746; *MR* **33** #6506.

J. Hammer, [Ham].

G. H. Hardy & E. M. Wright, *Introduction to the Theory of Numbers*, 4th ed. Chapter 24, Oxford University Press, Oxford, 1960.

H. Minkowski, *Geometrie des Zahlen*, Liepzig, Berlin, 1896.

P. R. Scott, An area-perimeter problem, *Amer. Math. Monthly* **81** (1974) 884–885.

P. R. Scott, An analogue of Minkowski's theorem in the plane, *J. London Math. Soc.* (2) **8** (1974) 647–651; *MR* **51** #4058.

P. R. Scott, On Minkowski's theorem, *Math. Mag.* **47** (1974) 277; *MR* **51** #4050.

P. R. Scott, Convex bodies and lattice points, *Math. Mag.* **48** (1975) 110–112; *MR* **51** #1610.

P. R. Scott, A new extension of Minkowski's theorem, *Bull. Austral. Math. Soc.* **18** (1978) 403–405; *MR* **58** #21952.

P. R. Scott, Two problems in the plane, *Amer. Math. Monthly* **89** (1982) 460–461.

P. R. Scott, On the area of planar convex sets containing many lattice points, *Bull. Austral. Math. Soc.* **35** (1987) 441–454; *MR* **88e**:52017.

P. R. Scott, Modifying Minkowski's theorem, *J. Number Theory* **29** (1988) 13–20; *MR* **89f**:11087.

E14. Positioning convex sets relative to discrete sets. Many of the problems of the last few sections concerning lattice points can be asked for more general sets of points scattered across the plane and subject to some density condition.

The following perplexing problem is due to Danzer. Let S be a point set in $\mathbb{R}^2$ of bounded density, that is, with the number of points in $S \cap B_r$ at most cr^2 for some constant c, where B_r is the disk center the origin and radius r. Do there always exist convex sets of arbitrarily large area not containing any points of S? This is so if S is a finite union of lattices, but it is false for certain S with logarithmically greater density, i.e., with (roughly) $cr^2 \ln r$ points of S in B_r.

In the other direction, suppose that $\lim_{r \to \infty}$ (number of points in $S \cap B_r)/\pi r^2 = 1$. Steinhaus asks whether every domain (convex, or more generally, measurable) with an area of at least n, has a congruent copy covering n points of S. Macbeath & Rogers showed that for any bounded domain K of area 1 not including the origin there exists a linear transformation of determinant 1 mapping K to a set disjoint from S.

R. P. Bambah & A. C. Woods, On a problem of Danzer, *Pacific J. Math.* **37** (1971) 295–301; *MR* **46** #2556.

L. Danzer, Problem 6, [Fen].

A. M. Macbeath & C. A. Rogers, A modified form of Siegel's mean value theorem, *Proc. Cambridge Philos. Soc.* **51** (1955) 565–576; *MR* **17**, 241.

F. Finite Sets of Points

This chapter considers geometric properties of finite sets of points in the plane, in $\mathbb{R}^d$, or, sometimes, on the surface of a sphere. For example, in a set of n points, what can be said about the set of distances between pairs of points, or the set of angles or areas defined by triples?

The name of Erdős is particularly associated with this area of combinatorial geometry, to which he has contributed greatly. Many of the problems are due to him, and for some of them he has offered substantial sums for solutions. He has published numerous survey articles and, together with Purdy, is writing a book on the subject.

A. Baker, B. Bollobás & A. Hajnal (eds.), A tribute to Paul Erdős, Cambridge University Press, Cambridge, 1990.

P. Erdős, *The Art of Counting, Selected Writings*, J. Spencer (ed.), MIT Press, 1973; *MR* **58** #27144.

P. Erdős, Some combinatorial problems in geometry, in *Geometry and Differential Geometry (Haifa, 1979)*, Lecture Notes in Math. 792, Springer, Berlin, 1980, 46–53; *MR* **82d**:51002.

P. Erdős, Some old and new problems in combinatorial geometry, in [RZ], 129–136; *MR* **87b**:52018.

P. Erdős, Problems and results in combinatorial geometry, in [GLMP], 1–11; *MR* **87g**:52001.

P. Erdős, Some combinatorial and metric problems in geometry, in [BF], 167–177.

P. Erdős, On some metric and combinatorial geometric problems, *Discrete Math.* **60** (1986) 147–153; *MR* **88f**:52011.

P. Erdős & G. Purdy, Some more combinatorial problems in the plane, to appear.

P. Erdős & G. Purdy, Some extremal problems in combinatorial geometry, to appear.

H. Hadwiger, H. Debrunner & V. Klee, [HDK].

W. Moser & J. Pach, [MP].

W. O. J. Moser, Problems on extermal properties of a finite set of points, in [GLMP], 52–64; *MR* **87g**:52003.

F1.   **Minimum number of distinct distances.** Let $f(n)$ denote the minimum number of distinct distances that can occur among the $\binom{n}{2}$ distances between n distinct points in the plane. Also, let $r(n)$ be the number of essentially different configurations for which this minimum is realized. For example, $f(3) = 1$, $f(4) = f(5) = 2$, and $f(6) = f(7) = 3$, with $r(3) = 1$, $r(4) = 3$, and $r(5) = 1$ (see Figure F1).

The first asymptotic estimate was due to Erdős, who showed that

$$cn^{1/2} < f(n) < cn/(\ln n)^{1/2}.$$

For the lower bound, let $\mathbf{x}$ be any of the n points. Either the other points are at least $(n - 1)^{1/2}$ distinct distances from $\mathbf{x}$, or there is a circle with center $\mathbf{x}$ on which at least $(n - 1)^{1/2}$ of the points lie, which clearly realize at least $\frac{1}{2}(n - 1)^{1/2}$ distances between them. The upper bound comes from points arranged on a square lattice. Erdős conjectures that $f(n) > cn^{1-\varepsilon}$ for each $\varepsilon > 0$, perhaps even that $f(n) = [1 + o(1)]cn/(\ln n)^{1/2}$, and offers \$500 for a proof or a disproof. More than 30 years ago, Leo Moser showed that $f(n) > cn^{2/3}$; recently this was improved to $f(n) > cn^{5/7}$ by Chung, to $f(n) > cn^{58/81 - \varepsilon}$ by Beck, and to $f(n) > cn^{4/5}$ by Szemerédi.

Are there values of $n > 5$ for which $r(n) = 1$? Does the set of points realizing the minimum always have a lattice structure? For example, is there always a line which contains $cn^{1/2}$, or even cn^{ε} of the points for all $\varepsilon > 0$? Are there $cn^{1/2}$ lines, or even $cn^{1-\varepsilon}$ lines which contain all of the points? In a configuration for which $f(n)$ is attained, are there always four points which determine only two distances, or even four points which determine three distinct distances?

Let $f_1(n)$ denote the minimum number of distances that can occur among n points in *general position* (i.e., no three on a line and no four on a circle).

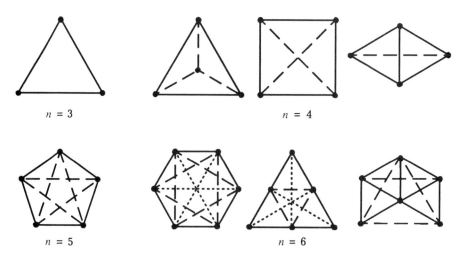

$n = 3$ $n = 4$

$n = 5$ $n = 6$

Figure F1.   Configurations of n points with the minimum possible number of distinct distances between pairs.

Erdős conjectured that $f_1(n)/n \to \infty$ and $f_1(n)/n^2 \to 0$ as $n \to \infty$. Recently, Erdős, Hickerson & Pach observed that the configuration obtained by taking a typical plane projection of the vertices of the unit cube in $\mathbb{R}^d$ gives $f_1(n) \le 4n^{\ln 3/\ln 2} = 4n^{1.585\cdots}$. What is the correct order of magnitude?

J. Beck, Different distances, to appear.

A. Blokhuis, Few-distance sets, *CWI Tract* 7, Stichting Math. Centrum, Amsterdam, 1984; *MR* **87f**:51023.

F. R. K. Chung, The number of different distances determined by n points in the plane, *J. Combin. Theory Ser. A* **36** (1984) 342–354; *MR* **86c**:52012.

P. Erdős, On sets of distances of n points, *Amer. Math. Monthly* **53** (1946) 248–250; *MR* **7**, 471.

P. Erdős, D. Hickerson & J. Pach, A problem of Leo Moser about repeated distances on the sphere, *Amer. Math. Monthly* **96** (1989) 569–575; *MR* **90h**:52008.

L. Moser, On different distances determined by n points, *Amer. Math. Monthly* **59** (1952) 85–91; *MR* **13**, 768.

E. Szemerédi, to appear.

F2. Repeated distances.

Let $g(n)$ denote the maximum number of times that the distance 1 (or, by scaling, any other distance) can occur between pairs of points from a set of n distinct points in the plane. Then $g(1) = 1$, $g(2) = 2$, $g(3) = 3$, $g(4) = 5$, $g(5) = 7$, $g(6) = 9$, and $g(7) = 12$. Years ago, Erdős showed that

$$n^{1+c/\ln \ln n} < g(n) < n^{3/2},$$

the lower bound coming from the points on a square lattice. He conjectured that $g(n) < cn^{1+\varepsilon}$ for some $c > 0$, for each $\varepsilon > 0$, offering \$500 for a proof or disproof. Beck & Spencer showed that $g(n) < cn^{13/9}$ and Spencer, Szemerédi & Trotter improved this to $g(n) < cn^{4/3}$, though this is still a long way from the conjectured estimate.

Erdős and L. Moser conjecture that if the n points are the vertices of a strictly convex polygon, then $g(n) < cn$ for some constant c. They can show that in these circumstances $g(3m + 1) \ge 5m$ for each natural number m and Füredi has obtained $g(n) < cn \ln n$.

This problem is also of interest in higher dimensions. With $g(n)$ now as the maximum number of unit distances among n points in $\mathbb{R}^3$, Erdős proved that

$$cn^{4/3} < g(n) < cn^{5/3}.$$

Successive improvements to this are due to Beck, Chung, and Clarkson et al., and it is now known that

$$cn^{4/3} \ln \ln n < g(n) < cn^{8/5}.$$

In four dimensions the situation is very different. H. Lenz pointed out that, if $P = \{(u_i, v_i, 0, 0) : 1 \le i \le \lceil \frac{1}{2}n \rceil\}$ and $Q = \{(0, 0, u_j, v_j) : 1 \le j \le \lceil \frac{1}{2}n \rceil\}$ where (u_i, v_i) are distinct solutions of $u^2 + v^2 = \frac{1}{2}$, then the distance between any point of P and any point of Q is 1. Thus $P \cup Q$ is a set of n or $n + 1$ points in $\mathbb{R}^4$ with more than $\frac{1}{4}n^2$ unit distances, giving the correct order of magnitude in the 4-dimensional case.

J. Beck, On the lattice property of the plane and some problems of Dirac, Motzkin, and Erdős in combinatorial geometry, *Combinatorica* **3** (1983) 281–297; *MR* **85j**:52013.

J. Beck & J. Spencer, Unit distances, *J. Combin. Theory Ser. A* **37** (1984) 231–238: *MR* **86a**:52015.

F. R. K. Chung, Sphere-and-point incidence relations in high dimensions with applications to unit distances and furthest neighbour pairs, *Discrete Comput. Geom.* **4** (1989) 183–190; *MR* **90g**:52011.

K. L. Clarkson, H. Edelsbrunner, L. J. Guibas, M. Sharir & E. Welzl, Combinatorial complexity bounds for arrangements, to appear.

P. Erdős, On sets of distances of *n* points, *Amer. Math. Monthly* **53** (1946) 248–250; *MR* **7**, 471.

P. Erdős, On sets of distances of *n* points in Euclidean space, *Magyar Tud. Akad. Mat. Kutató Int. Közl.* **5** (1960) 165–168; *MR* **25** #4420.

Z. Füredi, The maximum number of unit distances in a convex *n*-gon, *J. Combin. Theory Ser. A* **55** (1990) 316–320.

S. Józsa & E. Szemerédi, The number of unit distance on the plane, in *Infinite and Finite Sets*, 939–950; Colloq. Math. Soc. J. Bolyai 10, North-Holland, Amsterdam, 1975, *MR* **52** #10624.

J. Spencer, E. Szemerédi & W. Trotter, Unit distances in the Euclidean plane, in *Graph Theory and Combinatorics*, Academic Press, London, 1984, 293–303; *MR* **86m**:52015.

F3.   Two-distance sets.

How many distinct points can one have in $\mathbb{R}^d$ with at most two different distances occuring between pairs? The answer is three in $\mathbb{R}^1$, five in $\mathbb{R}^2$ (forming a regular pentagon), and six in $\mathbb{R}^3$ (the vertices of a regular octahedron, or of a regular triangular prism, or of a regular pentagon together with a suitable point on its axis). The set of $\frac{1}{2}d(d+1)$ mid-points of the edges of a regular simplex always forms a two-distance set, and Delsarte, Goethals & Seidel give examples with $\frac{1}{2}d(d+3)$ points for $d = 2, 6$, and 22. The best general upper bound is due to Blokhuis who has shown that a two-distance set in $\mathbb{R}^d$ has at most $\frac{1}{2}(d+1)(d+2)$ points, but the exact maximum is unknown for $d \geq 4$.

More generally, let $g_d(n)$ be the maximum number of points in $\mathbb{R}^d$ that determine at most *n* distances. What is $g_3(n)$ for small *n*? In particular, is the only 12-point, 3-distance configuration in $\mathbb{R}^3$ the set of vertices of an octahedron? And, presumably harder, is the only 20-point, 5-distance set the set of vertices of a dodecahedron?

Bannai, Bannai & Stanton and Blokhuis have shown that $g_d(n) \leq \binom{d+n}{n}$. Is $g_d(n) \sim d^n$ the correct order of magnitude for large *n*?

E. Bannai & E. Bannai, An upper bound for the cardinality of an *s*-distance subset in real Euclidean space, *Combinatorica* **1** (1981) 99–102; *MR* **82k**:05030.

E. Bannai, E. Bannai & D. Stanton, An upper bound for the cardinality of an *s*-distance subset in real Euclidean space II, *Combinatorica* **3** (1983) 147–152; *MR* **85e**:52013.

A. Blokhuis, An upper bound for the cardinality of *s*-distance sets in E^d and H^d, Eindhoven Univ. Tech. Mem. 68 (1982).

A. Blokhuis, A new upper bound for the cardinality of 2-distance sets in Euclidean space, in [RZ], 65–66; *MR* **86h**:52008.

H. T. Croft, 9-point and 7-point configurations in 3-space, *Proc. London Math. Soc.* (3) **12** (1962) 400–424, corrigendum **13** (1963) 384; *MR* **27** #5167a,b.

P. Delsarte, J. M. Goethals & J. J. Seidel, Spherical codes and designs, *Geom. Dedicata*
  6 (1977) 363–388; *MR* **58** #5302.
L. M. Kelly, Isosceles *n*-points, *Amer. Math. Monthly* **54** (1947) 227–229.
D. G. Larman, C. A. Rogers & J. J. Seidel, On two-distance sets in Euclidean space,
  Bull. London Math. Soc. **9** (1977) 261–267; *MR* **56** #16511.

F4.   Can each distance occur a different number of times? Erdős asked if it
is possible to have *n* points in general position in the plane (no three on a
line or four on a circle) such that for every i $(1 \leq i \leq n - 1)$ there is a distance
determined by the points that occurs exactly *i* times. (Note that $1 + 2 + \cdots +$
$(n - 1) = \binom{n}{2}$ = number of distinct pairs of points.) It seems certain that this
is not possible if *n* is large enough, but Erdős offers $50 for a proof (and $500
for examples with *n* arbitrarily large).

So far examples have been discovered for $2 \leq n \leq 8$; these are due to
Pomerance (5), several Hungarian students (6), and Palásti (7, 8); this last case
is illustrated in Figure F2. Can the construction be proved impossible for
$n = 9$, or is there an example?

Is it possible to have *n* arbitraily large for the same problem in $\mathbb{R}^3$, or,
indeed, in $\mathbb{R}^d$ for any $d \geq 2$?

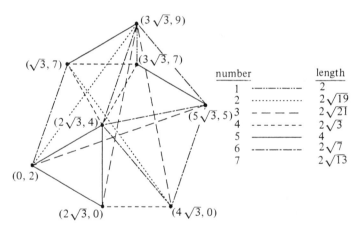

Figure F2.   Eight points in the plane arranged so that the *r*th distance is realized
between *r* pairs of points, for $r = 1, 2, \ldots, 7$.

P. Erdős, Distances with specified multiplicities, *Amer. Math. Monthly* **96** (1989) 447.
A. Liu, On the "seven points problem" of P. Erdős, *Math. Chronicle* **15** (1986) 29–33;
  MR **88e**:52014.
I. Palásti, On the seven points problem of P. Erdős, *Studia Sci. Math. Hungar.* **22** (1987)
  447–448; *MR* **89d**:52024.
I. Palásti, A distance problem of P. Erdős with some further restrictions, *Discrete Math.*
  76 (1989) 155–156.
I. Palásti, Lattice point examples for a question of Erdős, *Period. Math. Hungar.* **20**
  (1989) 231–235; *MR* **90j**:52012.

F5.   Well-spaced sets of points. Let S be a set of points in $\mathbb{R}^d$ with minimum mutual distance 1. Let

$$f_d(n) = \min_{|S|=n} D(S),$$

where $D(S)$ is the diameter of the set S (i.e. the maximum distance between any two points). The exact value of $f_d(n)$ is known for very few values of d and n. A number theoretic result of Thue states that

$$\lim_{n\to\infty} f_2(n)/n^{1/2} = 2^{1/2}3^{1/4}\pi^{-1/2}.$$

However, the value of $\lim_{n\to\infty} f_3(n)/n^{1/3}$ remains an unsolved problem in the geometry of numbers. These problems are related to optimal packing densities of balls (see Section D10).

Suppose now that S is a set of n points in the plane with minimum mutual distance 1 and, in addition, any two distances that are unequal differ by at least 1. Prove that the diameter of the set is at least cn where c is a constant independent of S. For example, the set of n equally spaced points on a straight line has diameter $n - 1$. Is it possible to do better? For $n = 7$, a regular hexagon and its center would only give $4 + 2\sqrt{3} = 7.464\ldots > 7 - 1$.

P. Erdős, Some combinatorial problems in geometry, in *Geometry and Differential Geometry (Haifa, 1979)* Lecture Notes in Math. 792, Springer, Berlin, 46–53, *MR* **82d**:51002.

P. Erdős, Aufgabe 856A, *Elem. Math* **36** (1981) 22.

F6.   Isosceles triangles determined by a set of points. How many isosceles triangles can occur with vertices in an n-point subset of the plane? According to Erdős & Purdy, at most $n(n - 1)$, but examples show that $(n - 2)(n - 4)$, or $(n - 1)(n - 2)$ if $n \equiv 0$ or 2 (mod 6), are possible.

Erdős also asked for the greatest number of points that can be placed in d-dimensional Euclidean space such that every triplet forms an isosceles triangle. The answer is 6 in the plane, the unique configuration being a regular pentagon and its center. In 3-space, if you adjoin two points to this configuration on its axis, at a distance from the center equal to the circumradius of the pentagon, you see that 8 points are possible, and Croft has proved that 9 are not. In 4 and 5 dimensions, Kelly has exhibited configurations of 10 and 16 points, respectively, but it is not known whether these are the largest possible.

The following problem appears in the collection by Moser & Pach. Show that given $\varepsilon > 0$, there is a constant c, such that from any set of n points in the plane one can always find a subset of $cn^{1-\varepsilon}$ points that contains the vertices of no isosceles triangle.

H. T. Croft, 9-point and 7-point configurations in 3-space, *Proc. London Math. Soc.* (3) **12** (1962) 400–424, corrigendum **13** (1963) 384; *MR* **27** #5167a,b.

P. Erdős, On some geometrical problems, *Math. Lapok* **8** (1957) 86–92; *MR* **20** #6056.

P. Erdős & G. Purdy, Some extremal problems in geometry IV, in *Proc. S. E. Conf. on Combinatorics, Graph Theory and Computing*, Congressus Numerantium XVII, Utilitas Math., 1976, 307–322; *MR* **55** #10292.
L. M. Kelly, Isosceles *n*-points, *Amer. Math. Monthly* **54** (1947) 227–229.
W. Moser & J. Pach, Problem 12, [MP].

F7. **Areas of triangles determined by a set of points.** The problems of Sections F2 and F1 may be posed with areas of triangles defined by triples of points replacing distances between pairs of points. What is the greatest number of different triangles of unit area (or any other constant positive area) that can occur with vertices in a set of *n* points in the plane? Erdős showed that

$$c_1 n^2 \ln \ln n < \text{max number of triangles of unit area} < c_2 n^{5/2}$$

for constants c_1 and c_2. The points (r, s) with integer coordinates satisfying $1 \le r \le a = \lfloor (\ln n)^{1/2} \rfloor$ and $1 \le s \le n/a$, give the left-hand estimate (with area $a!/2$). Erdős and Purdy have studied this problem in higher dimensions.

We can also ask for the minimum number of different areas of triangles that can occur with vertices among *n* points in the plane, not all on one straight line. It is known that

$$0.32n - 1 \le \text{minimum number of different areas} \le \lfloor \tfrac{1}{2}(n - 1) \rfloor;$$

the left-hand estimate is due to Burton & Purdy and the right-hand one is due to Erdős, Purdy & Straus. They conjecture that the right-hand value is the correct one, at least for sufficiently large *n*.

Write $r(S)$ for the ratio of the maximum area to the minimum area of the non-degenerate triangles with vertices in a plane set of points S, and let $f(n) = \inf_{|S|=n} r(S)$. Erdős, Purdy & Straus showed that $f(n) = \lfloor \tfrac{1}{2}(n - 1) \rfloor$ for $n > 37$, and conjecture that equality holds for $n > 5$. This value is achieved by sets of equally spaced points on a pair of parallel lines.

G. R. Burton & G. B. Purdy, The directions determined by *n* points in the plane, *J. London Math. Soc.* **20** (1979) 109–114; *MR* **81d**:52006.
P. Erdős & G. B. Purdy, Some extremal problems in geometry, *J. Combin. Theory* **10** (1971) 246–252.
P. Erdős & G. B. Purdy, Some extermal problems in geometry IV, in *Proc. Seventh S. E. Conf. on Combinatorics, Graph Theory and Computing, Baton Rouge*, Congressus Numerantium, XVII, Utilitas Math., Winnipeg, 1976, 307–322; *MR* **55** #10292.
P. Erdős, G. B. Purdy & E. G. Straus, On a problem in combinatorial geometry, *Discrete Math.* **40** (1982) 45–52.

F8. **Convex polygons determined by a set of points.** From any set of five points in the plane, no three collinear, it is possible to select four that are the vertices of a convex quadrilateral. This was shown by Ester Klein, who asked what the least integer $f(n)$ is such that from any $f(n)$ points in the plane, no three collinear, one can always select the vertices of a convex *n*-gon.

Erdős & Szekeres showed that

$$2^{n-2} + 1 \le f(n) \le \binom{2n-4}{n-2} + 1$$

and conjecture that $2^{n-2} + 1$ is the correct value. It is known that $f(4) = 5$ (Klein), $f(5) = 9$ (Kalbfleisch, Kalbfleisch & Stanton), and the first unsettled case is whether $f(6) = 17$.

More recently, Erdős asked what the least integer $g(n)$ is such that from any $g(n)$ points in the plane, no three collinear, one can select the vertices of a convex n-gon with none of the other points in its interior. It is easy to see that $g(3) = 3$ and $g(4) = 5$, and Harborth showed that $g(5) = 10$. However, Horton showed that there are arbitrarily large sets of points with no "empty" convex heptagons, so that $g(7)$ does not exist. Thus the outstanding question is whether $g(6)$ exists, and if so what its value is.

In a similar vein, given a set of k points, estimate how many distinct convex n-gons and empty convex n-gons are determined.

I. Bárány & Z. Füredi, Empty simplices in Euclidean space, *Canad. Math. Bull.* **30** (1987) 436–445; *MR* **89g**:52004.

P. Erdős, Some combinatorial problems in geometry, in *Geometry and Differential Geometry (Haifa, 1979)*, Lecture Notes in Math. 792, Springer, Berlin, 1980, 46–53; *MR* **82d**:51002.

P. Erdős & G. Szekeres, A combinatorial problem in geometry, *Composito Math.* **2** (1935) 263–270.

P. Erdős & G. Szekeres, On some extremum problems in elementary geometry, *Ann. Univ. Sci. Budapest* **3–4** (1960–61) 53–62; *MR* **24** A #3560.

H. Harborth, Konvexe Fünfecke in ebenen Punktmengen, *Elem. Math.* **33** (1978) 116–118; *MR* **80a**:52003.

J. D. Horton, Sets with no empty convex 7-gons, *Canad. Math. Bull.* **26** (1983) 482–484; *MR* **85f**:52007.

S. Johnson, A new proof of the Erdős–Szekeres convex k-gon result, *J. Combin. Theory Ser. A* **42** (1986) 318–319; *MR* **87j**:52004.

J. D. Kalbfleisch, J. G. Kalbfleisch & R. G. Stanton, A combinatorial problem on convex n-gons, in *Proc. Louisiana Conf. on Combinatorics, Graph Theory and Computing*, Louisiana State University, Baton Rouge, 1970, 180–188; *MR* **42** #8394.

M. Katchalski & A. Meir, On empty triangles determined by points in the plane, *Acta Math. Hungar.* **51** (1988) 323–328; *MR* **89f**:52021.

F9. Circles through point sets.

Every three non-collinear points in the plane determine a circle. Erdős asks for the maximum number $f(n)$ of congruent circles that can be so determined by a set of n points in the plane. Since every pair of points lies on at most two unit circles, and counting such pairs of points counts each circle at least three times, it follows that $f(n) \le \frac{2}{3}\binom{n}{2} = \frac{1}{3}n(n-1)$. On the other hand, Elekes gave a neat construction to show that $f(n) > cn^{3/2}$ for a constant c. What is the exact order of magnitude of $f(n)$? For small values of n, Harborth (& Mengersen) show that $f(3) = 1$, $f(4) = f(5) = 4$, $f(6) = 8$, $f(7) = 12$, and $f(8) = 16$. What are the next few values?

G. Elekes, *n* points in the plane can determine $n^{3/2}$ unit circles, *Combinatorica* **4** (1984) 131; *MR* **86e**:52015.

P. Erdős, Some applications of graph theory and combinatorical theory to number theory and geometry, *Colloq. Math. Soc. Janos Bolyai* **25** (1981) 137–148; *MR* **83g**:05001.

H. Harborth, Einheitskreise in ebenen Punktmengen, 3, in *Kolloquium über Diskrete Geometric*, Inst. Math. Univ. Salzburg, Salzburg, 1985, 163–168.

H. Harborth & I. Mengersen, Points with many unit circles, *Discrete Math.* **60** (1986) 193–197; *MR* **87m**:52015.

F10. Perpendicular bisectors.

Some years ago, Croft asked if it was possible to have a finite set of points in the plane with the property that the perpendicular bisector of any pair of them passes through at least two other points of the set. Kelly has found such a set of eight points (see Figure F3). This set consists of the vertices of a square together with the vertices of four equilateral triangles erected outwards (or inwards, with a similar result!) on the sides of the square. What other sets of points have this property? Is eight the largest number of points for which this is possible? Are there any configurations with at least three points on every perpendicular bisector? Which point sets in space have such properties (where "perpendicular bisectors" are now planes)?

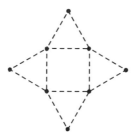

Figure F3. Eight points such that the perpendicular bisector of any pair passes through two other points.

F11. Sets cut off by straight lines.

The following problem is due to G. J. Simmons. Given an even number *n* of points in the plane, no three on a line, we call a line through a pair of them a **bisector** if it divides the plane into two parts with $\frac{1}{2}n - 1$ points in each part. Estimate $f(n)$, the maximum number of bisectors that any configuration can have. Straus used the following splitting process to show that $f(n) \geq n \log_2 n$. Given a configuration of points, choose a bisector through each point with different directions. Split each point into a close pair in the directions chosen, then you get two new bisectors for each pair together with all the previous ones. Iterating this procedure gives Straus' estimate. It seems that $n \log_2 n$ should be the correct order of magnitude, but all that is known at the moment is that $f(n) \leq cn^{3/2}$ for a suitable constant c.

More generally, if $1 \leq k \leq \frac{1}{2}n$, we denote by $f_k(n)$ the greatest number of different k-point subsets that may be partitioned off from an n-point set by a straight line. Erdős et al., and Edelsbrunner & Welzl show that

$$c_1 n \ln(k + 1) \leq f_k(n) \leq c_2 n\, k^{1/2}.$$

Is $n \ln(k + 1)$ the correct order of magnitude?

The 3-dimensional analog of this problem has recently attracted some interest: how many essentially different ways can there be of dividing a set of n points in $\mathbb{R}^3$ (in general position) into two equal halves by a hyperplane? The best result to date is at most $O(n^{8/3+\varepsilon})$ for each $\varepsilon > 0$, due to Chazelle et al.

N. Alon & E. Györi, The number of small semispaces of a finite set of points in the plane, *J. Combin. Theory. Ser. A* **41** (1986) 154–157; *MR* **87f**:52014.

I. Bárány & D. G. Larman, A coloured version of Tverberg's theorem, to appear.

I. Bárány, Z. Füredi & L. Lovász, On the number of halving planes, to appear.

B. Chazelle, H. Edelsbrunner, L. J. Guibas & M. Sharir, Points and triangles in the plane and halving planes in space, to appear.

H. Edelsbrunner & E. Welzl, On the number of line separations of a finite set in the plane, *J. Combin. Theory Ser. A* **38** (1985) 15–29; *MR* **86m**:52012.

P. Erdős, L. Lovász, A. Simmons & E. G. Straus, Dissection graphs of planar point sets, in *A Survey of Combinatorial Theory*, North-Holland, Amsterdam, 1973, 139–149; *MR* **51** #241.

J. E. Goodman & R. Pollock, On the number of k subsets of n points in the plane, *J. Combin. Theory Ser A* **36** (1984) 101–104; *MR* **85d**:05015.

G. Stöckl, Gesammelte und neue Eergebnisse über extreme k-Mengen für ebene punktmengen, Diplomarbeit, Inst. fur Information Processing, Technical University of Graz, 1984.

E. Welzl, More on k-sets of finite sets in the plane, *Discrete Comput. Geom.* **1** (1986) 95–100; *MR* **87f**:52015.

F12. Lines through sets of points.

Let X be a set of n points in the plane, not all lying on one line. At least n different lines are obtained by drawing the straight lines passing through two or more of the points of X. To see this, assume inductively that the result holds for $n - 1$ points. Then, given n points, take a line L that passes through exactly two of the points (see Sylvester's problem below) and delete one of these points. This leaves $n - 1$ points; either they are collinear, in which case the set consists of $n - 1$ collinear points and one other giving n lines, or, by induction, they determine $n - 1$ lines, to which may be added the line L.

A much more difficult problem is how many of the lines determined by n (not all collinear) points pass through some common point. Dirac conjectured that we can always find one of the n points that lies on at least $\frac{1}{2}n$ connecting lines. However, Grünbaum has catalogued six cases (with $n = 9, 15, 19, 25, 31,$ and 37) where this is false. Are these the only exceptions, and, in particular, is Dirac's conjecture true for all sufficiently large n? Beck, and independently Szemerédi & Trotter, have shown that there is a constant $c > 0$ such that, given n non-collinear points, there is always a point with at least cn distinct connecting lines through it, but their value of c is minute. Any substantial increase in the size of c would be of interest, since such results have considerable consequences in discrete combinatorial geometry.

In 1893, J. J. Sylvester published the following problem in the Educational Times. Given n points in the plane, not all on one line, show that there exists a straight line passing through *exactly two* of the points. While no correct solution was published at the time, the problem was revived by Erdős in the 1930s, and a number of solutions were then given.

The nicest argument is due to L. M. Kelly. Given a set X of n points, let $\mathbf{x}$ be the point and L the line passing through at least two of the points of X such the perpendicular distance from $\mathbf{x}$ to L has the least possible positive value, h, say. We claim that L contains just two points. Otherwise, if the foot of the perpendicular from $\mathbf{x}$ to L meets L at $\mathbf{z}$, at least two points of X, say $\mathbf{y}_1$ and $\mathbf{y}_2$, are on L on the same side of $\mathbf{z}$, taking $\mathbf{y}_1, \mathbf{y}_2, \mathbf{z}$ in that order on L. But now it is easy to see that the perpendicular distance from $\mathbf{y}_2$ to the line through $\mathbf{y}_1$ and $\mathbf{x}$ must be strictly less than h, giving a contradiction.

This problem has many variants and generalizations, and has an enormous literature devoted to it, a selection of which is listed at the end of the section. We now mention a few of the problems that remain unsolved.

We call a line that contains exactly two points from a given set an **ordinary line** or a **Gallai line**. Given n points in the plane, not all on one line, we denote by t the number of ordinary lines determined by the points. What is the minimum of t, over all configurations of n points? From the solution to Sylvester's problem, the minimum is at least one, and for $n \le 14$ the following values are known:

n	3	4	5	6	7	8	9	10	11	12	13	14
$t \ge$	3	3	4	3	3	4	6	5	6	6	6	7

Motzkin, and also Dirac, conjectured that $t \ge \lfloor \frac{1}{2}n \rfloor$ for any configuration. However, if n is even there exist configurations with $t = \frac{1}{2}n$. Kelly & Moser showed that $t \ge 3n/7$, and Hansen that $t \ge \frac{1}{2}n$, except for $n = 7$ and 13. But Hansen's argument is long and difficult to follow. Is there a shorter argument? What is the minimum value of t for odd $n \ge 15$?

Another famous problem, also due to Sylvester, has become known as the "Orchard Problem." Given n points in the plane, no four on a line, how many lines of three points can there be? There are constructions due to Burr and Furedi & Palasti with $\frac{1}{6}n^2 + O(n)$ lines of three points; is $\frac{1}{6}$ the largest constant possible?

Similarly, given n points, no five on a line, how many lines of four can there be? Grünbaum gave an example with $cn^{3/2}$ lines of four; Erdős claims that this is the right order of magnitude, and offers $250 for a proof, or $100 for the weaker conjecture that there are at most $o(n^2)$ lines of four.

Obviously, these problems have analogues in three dimensions, where we can count planes determined by sets of points, as well as lines. Suppose that n points in $\mathbb{R}^3$ are not all on one plane. Then if the points determine p distinct planes and l distinct lines, it is easy to see that $p \geq n$ and $l \geq n$. Is it true that for all sufficiently large n, perhaps even for $n \geq 17$, we have $p \geq l$, unless either the points lie on a pair of skew lines, or all but one of the points lie on a plane? Purdy has shown that except in these cases, $p \geq cl$ for some $c > 0$, and Purdy & Erdős have shown that $p \geq t + 2 - n$ if no three points are collinear and n is sufficiently large. Welsh asks if $l^2 \geq pn$ except in degenerate cases. Such results would have importance in matroid theory.

Of course, the most general problem is to determine exactly which triples (n, l, p) can occur for any configurations of points in $\mathbb{R}^3$.

A more detailed account of these topics is given by Erdős & Purdy.

J. Beck, On the lattice property of the plane and some open problems of Dirac, Motzkin, and Erdős in combinatorial geometry, *Combinatorica* **3** (1983) 281–297; *MR* **85j**:52013.

S. Burr, Planting trees, in [Kla], 90–99.

S. A. Burr, B. Grünbaum & N. J. A. Sloane, The orchard problem, *Geom. Dedicata* **2** (1974) 397–424; *MR* **49** #2428.

D. W. Crowe & T. A. McKee, On Sylvester's problem for collinear points, *Math. Mag.* **41** (1968) 30–34; *MR* **38** #3761.

G. A. Dirac, Collinearity properties of sets of points, *Quart. J. Math. Oxford Ser.* (2) **2** (1951) 221–227; *MR* **13**, 270.

P. Erdős & G. Purdy, Some extremal problems in combinatorial geometry, to appear.

Z. Füredi & I. Palásti, Arrangements of lines with a large number of triangles, *Proc. Amer. Math. Soc.* **92** (1984) 561–566; *MR* **86i**:52005.

B. Grünbaum, Arrangements and spreads, *Amer. Math. Soc. Providence* (1972); *MR* **46** #6148.

B. Grünbaum, New views on some old questions of combinatorial geometry, *Proc. Colloq. Int. sulle Teorie Combinatorie*, Acad. Nazionale die Lincei, 1976, Tomo I, 451–468; *MR* **57** #10605.

S. Hansen, Contributions to the Sylvester–Gallai theory, Dissertation, University of Copenhagen, 1981.

L. M. Kelly & W. O. J. Moser, On the number of ordinary lines determined by n points, *Canad. J. Math.* **10** (1958) 210–219; *MR* **20** #3494.

T. Motzkin, The lines and planes connecting the points of a finite set, *Trans. Amer. Math. Soc.* **70** (1951) 451–464; *MR* **12**, 849.

G. Purdy, Two results about points, lines, and planes, *Discrete Math.* **60** (1986) 215–218; *MR* **88a**:52011.

G. Purdy & P. Erdős, Some external problems in geometry III, *Congressus Numerantium XIV* (1975) 291–308; *MR* **52** #13650.

E. Szemerédi & W. T. Trotter, External problems in discrete geometry, *Combinatorica* **3** (1983) 381–392; *MR* **85j**:52014.

D. J. A. Welsh, *Matroid Theory*, Academic Press, London, 1976; *MR* **55** #148.

F13. Angles determined by a set of points.

If a set of n points in the plane do not all lie on one line, must three of them determine a (non-zero) angle less than π/n? (This is trivial if no three points are collinear.) We mentioned in Section F12 that, in a set of n points, there is always a point lying on at least cn of the connecting lines defined by the points, from which it follows that some three points determine an angle of at most c_1/n, but the best known constant c_1 is minute.

In the other direction, Erdős & Szekeres showed that there is always an angle of at least $\pi(1 - 1/\log_2 n) + O[(\log_2 n)^{-3}]$.

A related question is whether n points, not all on a line, determine at least $n - 2$ different angles. Again, the result of Section F12 above implies at least cn different angles are subtended at some point, for a very small value of c.

Suppose, now, that we are given a set S of n points in the plane, no three collinear, so that $N = \frac{1}{2}n(n - 1)(n - 2)$ angles are determined. In order to study the distribution of angles, we write $f(S, \alpha)$ and $g(S, \alpha)$ for the number of angles determined by S that are greater than α and less than α, respectively, for $0 < \alpha < \pi$. Let

$$f(n, \alpha) = \min_{|S|=n} f(S, \alpha), \qquad g(n, \alpha) = \min_{|S|=n} g(S, \alpha)$$

and

$$F(\alpha) = \lim_{n \to \infty} f(n, \alpha)/N, \qquad G(\alpha) = \lim_{n \to \infty} g(n, \alpha)/N.$$

Conway, Croft, Erdős & Guy show that these limits exist and that $F(\alpha)$ increases and $G(\alpha)$ decreases with α. Moreover, $F(\alpha) = \frac{1}{3}$ $(0 < \alpha \le \frac{1}{3}\pi)$, $G(\alpha) = \frac{2}{3}$ $(\alpha > \frac{1}{2}\pi)$, and $\frac{1}{9} < F(\frac{1}{2}\pi) \le \frac{4}{27}$. The function F is discontinuous at $\frac{1}{3}\pi$. Determine more values and properties of F and G, in particular at which α they are discontinuous, and what the exact value of $F(\frac{1}{2}\pi)$ is.

These problems have obvious analogs in three or more dimensions.

I. Bárány, An extension of the Erdős–Szekeres theorem on large angles, *Combinatorica* **7** (1987) 161–169; *MR* **88m**:52013.

I. Bárány & J. Lehel, Covering with Euclidean boxes, *European J. Combin.* **8** (1987) 113–119; *MR* **88i**:52014.

J. H. Conway, H. T. Croft, P. Erdős & M. J. T. Guy, On the distribution of values of angles determined by coplanar points, *J. London Math. Soc* (2) **19** (1979) 137–143; *MR* **80h**:51021.

P. Erdős, Some old and new problems in combinatorial geometry, in [RZ], 129–136; *MR* **87b**:52018.

P. Erdős & Z. Füredi, The greatest angle among n points in d-dimensional Euclidean space, in *Combinatorial Mathematics*, North-Holland Math. Studies 75, North-Holland, Amsterdam, 1983, 275–283; *MR* **87g**:52018.

P. Erdős & G. Szekeres, On some extremum problems in elementary geometry, *Ann. Univ. Sci. Budapest Eötvös Sect. Math.* **III/IV** (1960/61) 53–62.

G. Purdy, Repeated angles in E_4, *Discrete Comput. Geom.* **3** (1988) 73–75; *MR* **89a**:52028.

F14. Further problems in discrete geometry. This section demonstrates how many further problems in discrete geometry may be generated.

Let τ denote some type of configuration of points (e.g., "equilateral triangles"). Then we may ask the following questions:

1. What is the greatest number $f(n, \tau)$ of subsets of type τ that can occur in an n-point subset of the plane?

2. What is the least number $g(n, \tau)$ such that, given any set of n points in

the plane, we can always remove some $g(n, \tau)$ of the points so that those remaining contain no subset of type τ?

To avoid trivial cases, we may, for certain τ, need to restrict attention to sets of n points satisfying some condition such as no three collinear or no four on a circle.

It is rare to get exact solutions to such problems for general n. Usually, the most we can hope to do is to find exact values of $f(n, \tau)$ and $g(n, \tau)$ and the corresponding extremal configurations for certain small values of n, and find asymptotic estimates for large n.

We have already met a number of problems of this class, for example where τ is "pairs of points at unit distance," "isosceles triangles," "triangles of unit area," or "triangles of unit circumradius." However, there are many more possibilities that might be studied, such as with τ "equilateral triangles," "vertices of a square," "vertices of a regular k-gon," "triples of collinear points," "triples of points, one of which is the mid-point of the others," etc.

In a similar way, let m_k denote some "measure" determined by k points (where k is fixed) (e.g., if $k = 2$, then m_2 might be "the distance between a pair of points"). Then we might ask the following question:

3. What is the least number $h(n, m_k)$ such that for every set of n points in the plane, the k-point subsets give rise to at least $h(n, m_k)$ different values of m_k?

Here we might have m_2 as "the direction determined by a pair of points," and m_3 might be "the area/perimeter/inradius/circumradius of a triangle," etc.

We can further pose problems such as these for specific types of sets, (e.g., n-point subsets of the square lattice). Many of the problems make sense on the surface of a sphere. Of course, all of these problems have analogs in $\mathbb{R}^d$ for $d \geq 3$, when further types of configurations, such as "vertices of a regular tetrahedron," become possible.

F15. The shortest path joining a set of points.

F15. **The shortest path joining a set of points.** Given n points in the unit square, there is a shortest polygonal curve connecting them. How long might this path be? Newman gives simple arguments to show that the maximum length required lies between $\sqrt{n-1}$ and $2\sqrt{n} + 4$, and Few shows that one can always manage with length at most $\sqrt{2n} + 7/4$. What is the least constant c such that a length $c\sqrt{n} + o(\sqrt{n})$ suffices? What is the minimum length required for small values of n? It is easily seen to be $\sqrt{2}$ for $n = 2$ and $\frac{1}{2}(2 + \sqrt{5})$ for $n = 3$.

A **spanning tree** of a finite set of points X in the plane is a connected network of line segments joining various pairs of points of X that connects X [see Figure F4(a)]. A **Steiner tree** of X is again a network of line segments that connects X, but this time the line segments may meet each other at points not in X [see Figure F4(b)]. A spanning tree, respectively, a Steiner tree, is **minimal** if the total length of the segments is as small as possible. We write $L(X)$ for the length of a minimal spanning tree of X and $L_s(X)$ for the length of a minimal Steiner tree of X. Clearly $L_s(X) \leq L(X)$, but with the improvements

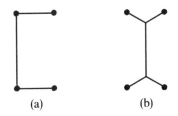

(a)   (b)

Figure F4.   (a) A minimal spanning tree and (b) the minimal Steiner tree through four points.

possible by allowing the segments to meet outside X, we get strict inequality in many cases.

Given an n-point set X in the unit square, obtain upper bounds for $L(X)$ and $L_s(X)$. Chung & Graham have shown that $L_s(X) \leq 0.995\sqrt{n} + O(1)$. Can the value 0.995 be reduced?

If X consists of the vertices of an equilateral triangle of unit side, then $L(X) = 2$ and $L_s(X) = \sqrt{3}$. Gilbert & Pollack conjectured that this case gives the least value of the ratio of $L_s(X)$ to $L(X)$, that is,

$$\rho = \inf_{x} \frac{L_s(X)}{L(X)} = \tfrac{1}{2}\sqrt{3} = 0.8660\ldots,$$

where the infimum is taken over all finite configurations of points X.

The conjecture is true for sets of up to five points, (see Du, Hwang & Yao or Friedel & Widmayer). A succession of improved lower bounds for ρ have been obtained in the general case, the best to date is $\rho \geq 0.8241$, due to Chung & Graham and requiring considerable computation.

Clearly, minimal trees are of importance in practical network problems, and quite a bit is known about their form. For example, in a Steiner minimal tree, the line segments meet in threes at points not in X, each making an angle of 120° with the other two.

F. Chung, M. Gardner & R. Graham, Steiner trees on a checkerboard, *Math. Mag.* **62** (1989) 83–96.

F. R. K. Chung & R. L. Graham, On Steiner trees for bounded sets of points, *Geom. Dedicata* **11** (1981) 353–361; *M R* **82j**:05072.

F. R. K. Chung & R. L. Graham, A new bound for Euclidean Steiner minimal trees, in [GLMP], 328–346; *M R* **87b**:05072.

F. R. K. Chung & F. K. Hwang, A lower bound for the Steiner tree problem, *SIAM J. Appl. Math.* **34** (1978) 27–36; *M R* **81h**:05052.

D. Z. Du & F. K. Hwang, A new bound for the Steiner ratio, *Trans. Amer. Math. Soc.* 278 (1983) 137–148; *M R* **85i**:05080a.

D. Z. Du, F. K. Hwang & E. Y. Yao, The Steiner ratio conjecture is true for five points, *J. Combin. Theory Ser. A* **38** (1985) 230–240; *M R* **86g**:05027.

L. Fejes Tóth, Über einen geometrischen Satz, *Math. Z.* **46** (1940) 83–85; *M R* **1**, 263.

L. Few, The shortest path and the shortest road through n points, *Mathematika* **2** (1955) 141–144; *M R* **17**, 1235.

J. Friedel & P. Widmayer, A simple proof of the Steiner ratio conjecture for five points, *SIAM J. Appl. Math.* **49** (1989) 960–967; *MR* **90h**:52009.

E. N. Gilbert & H. O. Pollak, Steiner minimal trees, *SIAM J. Appl. Math.* **16** (1968) 1–29; *MR* **36** #6317.

R. L. Graham & F. K. Hwang, Remarks on Steiner minimal trees, *Bull. Inst. Math. Acad. Sinica* **4** (1976) 177–182; *MR* **55** #10302.

R. Kallman, On a conjecture of Gilbert and Pollak on minimal trees, *Stud. Appl. Math.* **52** (1973) 141–151; *MR* **49** #7166.

Z. Melzak, On a problem of Steiner, *Canad. Math. Bull.* **4** (1961) 143–148; *MR* **23A** #2767.

D. J. Newman, *A Problem Seminar*, Problem 73, Springer, New York, 1982.

H. O. Pollak, Some remarks on the Steiner problem, *J. Combin. Theory Ser. A* **24** (1978) 278–295; *MR* **58** #10543.

S. Verblunsky, On the shortest path through a number of points, *Proc. Amer. Math. Soc.* **2** (1951) 904–913; *MR* **13**, 577.

F16. Connecting points by arcs.

John Conway has invented the following concept: A **thrackle** is a planar configuration consisting of n points, and m (smooth) arcs, each arc joining some distinct pair of the points, the arcs further satisfying the following conditions:

(1) every pair of them intersects *just* once, in a sense to be made precise below;

(2) no arc intersects itself;

where "intersect" is to be interpreted as *either* to cut, at a relatively interior point of each arc, and to cross over at this common point, *or* to have one common end point (but not both of these possibilities).

The Thrackle conjecture is that $m \leq n$. Some examples with equality are displayed (see Figure F5). "Experimental" evidence for $n \leq 10$ supports this conjecture, but a general proof is elusive. Is there at least a proof for small n?

Woodall, in an elegant paper, has succeeded in characterizing all thrackles, *if we assume the thrackle conjecture* and also in characterizing all **straight-line thrackles** (where a straight-line thrackle has all its arcs straight).

We may consider analogous constructions on other surfaces, and presumably expect (with obvious notation) the appropriate conjecture to be that $\max(m - n)$ depends on the genus of the surface.

Robert Connelly considers a problem concerning the lengths rather than the arrangements of the arcs. Let $\mathbf{x}_i$, $\mathbf{y}_i$, with $1 \leq i \leq n$, be n given pairs of

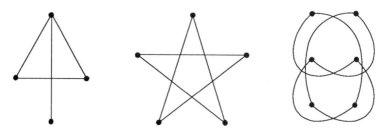

Figure F5.   Example of thrackles on 4, 5, and 6 vertices.

distinct points such that $|\mathbf{x}_i - \mathbf{y}_i| \leq 1$. Let L_i be the length of an arc A_i, joining $\mathbf{x}_i$ to $\mathbf{y}_i$, with all the A_i disjoint. The largest L_i, for fixed $\mathbf{x}_i$, $\mathbf{y}_i$, must be greater than some number $c(\mathbf{x}_i, \ldots, \mathbf{x}_n, \mathbf{y}_i, \ldots, \mathbf{y}_n)$, where c is chosen as small as possible. What is the maximum of c over all such sets $\{\mathbf{x}_i, \mathbf{y}_i\}$? The answers for $n = 1, 2$, and 3 may be $1, \sqrt{2}$ and $(\sqrt{3} + 1)/2$. If the $\mathbf{x}_i$, $\mathbf{y}_i$ are all in the unit square, say, what is the asymptotic behavior of this maximum c, as n becomes large?

Obviously one can ask many other problems that concern the topological and metrical realizations of graphs (i.e., configurations of points, some pairs of which are joined) in the plane.

D. R. Woodall, Thrackles and deadlock, *Combinatorial Math. and its Applications*, D. Welsh (ed.), Academic Press, New York, 1971, 335–347; *MR* **43** #6111.

F17. Arranging points on a sphere.

Many problems involve finding the configurations of n points on the surface of a unit sphere that maximize or minimize some given quantity (e.g., the sum of the distances between pairs of points). Of course, we wish to know whether the extremal configurations are unique (up to rotation and reflection). Also of interest are local maxima or minima, that is, configurations for which every small perturbation reduces or increases the quantity. Sometimes, but certainly not always, there is just one local extremum which must therefore be the global one.

In general, the most we can hope to do is to find exact solutions for small n, though there are sometimes a few rather larger values where the problem can be solved using certain highly symmetric configurations of points. However, even when explicit solutions cannot be found, it is often possible to make qualitative statements about optimal configurations.

Some of these questions are directly relevant to physics or chemistry, where stable configurations tend to minimize some form of energy expression.

All of these problems can be posed on other convex surfaces. Generally, difficulties of describing the geometry of such surfaces make exact solutions out of the question, but, again, qualitative descriptions may be possible.

(1) *The electron problem.* Each of the n points on the unit sphere is the position of a unit electric charge, and it is required to minimize the energy (i.e., to minimize the sum of the reciprocals of the distances between the points). Föppl gave answers for $n = 5, 6$, and 7 (double pyramids) and for 8, 10, 12, and 14 (North and South poles with an antiprism between). Cohn computed results for $n = 9$ and 11. His configurations consist of two equilateral triangles with an equatorial equilateral triangle, respectively, a certain non-regular pentagon, between.

(2) *The packing problem.* Maximize the least distance between any pair of points, or equivalently, pack n equal circles, or spherical caps, of largest possible radius, on the surface of the sphere. This is Tammes' problem (Section D7).

(3) *The covering problem.* Arrange n points on the sphere so that the maximum distance from anywhere on the sphere to the nearest point is as small as possible. Equivalently, minimize the radius of n equal circles, or spherical caps, which cover the sphere. This is the covering dual of the packing problem (2) above, and is discussed in Section D8.

(4) *Maximum average distance.* Maximize the sum of the $\binom{n}{2}$ distances between the n points. This is equivalent to maximizing the average distance. The problem is open for $n > 4$, but Berman & Hanes prove a lemma and use it to implement an iterative program yielding

Number of Points	5	6	7	8	9	10
Sum of Distribution	15.6814	22.9706	31.5309	41.4731	52.7436	65.3497
Average Distribution	1.56814	1.53137	1.50147	1.48118	1.46510	1.45222

The first three of these correspond to the points being at the vertices of a double pyramid. For large n, it is easy to see that the maximum distance sum is approximately cn^2, for a constant c corresponding to a "continuous uniform distribution" of points. It is known that the maximum obtainable differs by at most $c_1 n^{1/2}$ from this for a constant c_1.

We can also ask to maximize the sum of the squares of the distances.

(5) *Maximum convex hull.* Find the polyhedron with n vertices on the unit sphere and maximum volume. This has been solved for $n \leq 9$ by Berman & Hanes, who show that in general the maximum polyhedron is simplicial (i.e., has all its faces triangular). For $n = 4$, the solution is the regular tetrahedron, and for $5 \leq n \leq 7$ it is the triangular, square, and pentagonal double pyramid, respectively, so in particular it is the regular octahedron for $n = 6$. For $n = 8$ it has four valence-4 vertices $\mathbf{p}_2, \mathbf{p}_3, \mathbf{p}_6,$ and $\mathbf{p}_7$ and four valence-5 vertices, $\mathbf{p}_1$, $\mathbf{p}_4, \mathbf{p}_5,$ and $\mathbf{p}_8$ where the coordinates of $\mathbf{p}_i$, $1 \leq i \leq 8$, are $(\sin 3\phi, 0, \cos 3\phi)$, $(\sin \phi, 0, \cos \phi), (-\sin \phi, 0, \cos \phi), (-\sin 3\phi, 0, \cos 3\phi), (0, -\sin 3\phi, -\cos 3\phi)$, $(0, -\sin \phi, -\cos \phi), (0, \sin \phi, -\cos \phi), (0, \sin 3\phi, -\cos 3\phi)$, where $\cos^2 \phi = (15 + \sqrt{145})/40$. The volume is $[(475 + 29\sqrt{145})/250]^{1/2} = 1.815\ldots$. Grace conjectures that the maximum volume is always realized by a **medial** polyhedron, that is one with vertices of valence m or $m + 1$ where $m \leq 6 - \frac{12}{n} < m + 1$ (or all of valence-5 in case $n = 12$, presumably the icosahedron).

The problem of finding the polyhedron with n vertices on the sphere that is of maximum width also seems challenging.

(6) *Minimum circumscribed polyhedron.* This is the dual of the previous problem. Arrange the n points so that the n tangent planes to the sphere at the points enclose a polyhedron of minimum volume. Goldberg made a conjecture about the extremals dual to that formulated by Grace (see previous problem).

R. Alexander, On the sum of distances between n points on a sphere, I, II, *Acta. Math. Acad. Sci. Hungar.* **23** (1972) 443–448; **29** (1977) 317–320; *MR* **47** #957, **56** #3747.

R. Alexander & K. B. Stolarsky, Extremal problems of distance geometry related to energy integrals, *Trans. Amer. Math. Soc.* **193** (1974) 1–31; *MR* **50** #3121.

J. Beck, Sums of distances between points on a sphere—an application of the theory of irregularities of distribution to discrete geometry, *Mathematika* **31** (1984) 33–41; *MR* **86d**:52004.

J. D. Berman & K. Hanes, Volumes of polyhedra inscribed in the unit sphere in E^3, *Math. Ann.* **188** (1970) 78–84; *MR* **42** #961.

J. D. Berman & K. Hanes, Optimizing the arrangement of points on the unit sphere, *Math. Comput.* **31** (1977) 1006–1008; *MR* **57** #17502.

G. Björk, Distribtuions of positive mass which maximize a certain generalized energy integral, *Ark. Math.* **3** (1956) 255–269; *MR* **17**, 1198.

R. Bowen & S. Fisk, Generation of triangulations of the sphere, *Math. Comput.* **21** (1967) 250–252.

G. D. Chakerian & M. S. Klamkin, Inequalities for sums of distances, *Amer. Math. Monthly* **80** (1973) 1009–1017; *MR* **48** #9558.

H. Cohn, Stability configurations of electrons on a sphere, *Math. Tables Aids Comput.* **10** (1956) 117–120; *MR* **18**, 356.

L. Fejes Tóth, On the sum of distances determined by a pointset, *Acta. Math. Acad. Sci. Hungar.* **7** (1956) 397–401; *MR* **21** #5937.

L. Föppl, Stabile anordmungen von Elektronen in Atom, *J. Reine Angew. Math.* **141** (1912) 251–302.

M. Goldberg, The isoperimetric problem for polyhedra, *Tôhoku Math. J.* **40** (1935) 226–236.

M. Goldberg, Stability configurations of electrons on a sphere, *Math. Comput.* **23** (1969) 785–786.

D. W. Grace, Search for largest polyhedra, *Math. Comput.* **17** (1963) 197–199.

G. Harman, Sums of distances between points on a sphere, *Int. J. Math. Sci.* **5** (1982) 707–719.

E. Hille, Some geometrical extremal problems, *J. Austral. Math. Soc.* **6** (1966) 122–128; *MR* **33** #6507.

T. W. Melnyk, O. Knop & W. R. Smith, Extremal arrangements of points and unit charges on a sphere: equilibrium configurations revisited, *Canad. J. Chem.* **55** (1977) 1745–1761.

K. B. Stolarsky, Sums of distances between points on a sphere I, II, *Proc. Amer. Math. Soc.* **35** (1972) 547–549; **41** (1973) 575–582; *MR* **46** #2555, **48** #12314.

K. B. Stolarsky, Spherical distributions of N points with maximal distance sums are well spaced, *Proc. Amer. Math. Soc.* **48** (1975) 203–206; *MR* **51** #1615.

G. General Geometric Problems

Most of the problems encountered so far have involved convex sets, or other sets with considerable intrinsic geometric structure. Nevertheless, one can study the geometry of much more general objects, for example, sets that are just closed or Lebesgue measurable, or even completely arbitrary. This chapter contains a selection of problems which, at least at first glance, are of such a general nature.

G1. Magic numbers. The following surprising result has become known as Gross's theorem. If E is any compact connected subset of $\mathbb{R}^d$, there is a unique number m, known as the **magic number** of E, with the following property: Given a positive integer n and points $\mathbf{x}_1, \mathbf{x}_2, \ldots, \mathbf{x}_n$ in E, there exists a point $\mathbf{y}$ of E such that the average distance of the $\mathbf{x}_i$ from $\mathbf{y}$ is m, that is,

$$\frac{1}{n} \sum_{i=1}^{n} |\mathbf{x}_i - \mathbf{y}| = m.$$

In fact this is true in any metric space, and Stadje has shown that such results even hold for any symmetric continuous distance functions on a wide class of connected topological spaces.

A simple argument of Szekeres (see Cleary, Morris & Yost) shows that the magic number of any *convex* body in $\mathbb{R}^d$ equals its circumradius. Thus questions of interest concern non-convex sets, and in particular the boundary curves of plane convex sets. The circumference of the circle of diameter one has magic number $2/\pi$ and the perimeter of the equilateral triangle of side 1 has magic number $(2 + \sqrt{3})/6$. Cleary, Morris & Yost give an expression for the magic number of any regular polygon, but is there a formula for more general polygons? What is the magic number of the perimeter of an ellipse in terms of the lengths of its axes? Computations suggest that the perimeter of

168

the Reuleaux triangle of width 1 has a magic number of about 0.668, but what is the exact value?

What is the possible range of values of m for a plane connected set E of diameter $D = \sup\{|\mathbf{x} - \mathbf{y}| : \mathbf{x}, \mathbf{y} \in E\}$? Certainly $\frac{1}{2}D \leq m$, with equality in a large number of cases. It is suggested that the maximum value of m is attained by the perimeter of the Reuleaux triangle, but this has yet to be proved. Nickolas & Yost have shown that $m/D \leq 4/\pi\sqrt{3} = 0.735 \ldots$ and this has been improved to $m/D \leq (2 + \sqrt{3})/3\sqrt{3} = 0.718 \ldots$. If E is a plane *convex* set then $\frac{1}{2} \leq m/D \leq 1/\sqrt{2} = 0.707 \ldots$, with both inequalities best possible.

A recent account of this work is given by Cleary, Morris & Yost, where these and further open problems may be found.

J. Cleary & S. A. Morris, Numerical geometry ... not numerical topology, *Math. Chronicle* **13** (1984) 47–58.

J. Cleary, S. A. Morris & D. Yost, Numerical geometry—numbers for shapes, *Amer. Math. Monthly* **93** (1986) 260–275.

O. Gross, The rendezvous value of a metric span, in *Advances in Game Theory*, Ann. Math. Studies 52, Princeton University, Princeton, 1964, 49–53.

S. A. Morris & P. Nickolas, On the average distance property of compact connected metric spaces, *Arch. Math. (Basel)* **40** (1983) 459–463; *M R* **85g**:54020.

P. Nickolas & D. T. Yost, On the average distance property for subsets of Euclidean space, *Arch. Math. (Basel)* **50** (1988) 380–384; *M R* **89d**:51026.

W. Stadje, A property of compact, connected spaces, *Arch. Math. (Basel)* **36** (1981) 275–280; *M R* **83e**:54028.

J. Strantzen, An average distance result in Euclidean n-space, *Bull. Australian Math. Soc.* **26** (1982) 321–330; *M R* **84e**:52015.

D. T. Yost, Average distances in compact connected spaces, *Bull. Austral. Math. Soc.* **26** (1982) 331–342; *M R* **84e**:52016.

G2. Metrically homogeneous sets.

A **metrically homogeneous** set S in $\mathbb{R}^d$ is a set such that for all $\mathbf{x}, \mathbf{y} \in S$, the sets of distances

$$\{|\mathbf{z} - \mathbf{x}| : \mathbf{z} \in S\} \quad \text{and} \quad \{|\mathbf{z} - \mathbf{y}| : \mathbf{z} \in S\}$$

are identical. Metrically homogeneous subsets of $\mathbb{R}^2$ have been characterized by Grünbaum & Kelly: If S is finite then it must be the vertex set of a regular, quasi-regular (a convex cyclic polygon with alternate sides equal), or an evenly truncated regular polygon. If S is compact and infinite, it must be a major arc of a curve of constant width. What is the exact characterization of the compact metrically homogeneous subsets of $\mathbb{R}^3$ and $\mathbb{R}^d$ for $d > 3$?

B. Grünbaum & L. M. Kelly, Metrically homogeneous sets, *Israel J. Math.* **6** (1968) 183–197, correction in **8** (1970) 93–95; *M R* **39** #6180, **42** #3679.

G3. Arcs with increasing chords.

A curve C has the **increasing chord property** if $|\mathbf{x}_2 - \mathbf{x}_3| \leq |\mathbf{x}_1 - \mathbf{x}_4|$ whenever $\mathbf{x}_1, \mathbf{x}_2, \mathbf{x}_3$, and $\mathbf{x}_4$ lie in that order on C (see Figure G1). Binmore asked whether there exists an absolute constant c such that

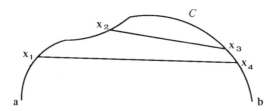

Figure G1. The increasing chord property. If x_1, x_2, x_3, and x_4 occur in that order on C, then $|x_2 - x_3| \leq |x_1 - x_4|$.

$$L \leq c|\mathbf{a} - \mathbf{b}|,$$

where C is a plane curve with the increasing chord property with length L and endpoints $\mathbf{a}$ and $\mathbf{b}$. Larman & McMullen showed that

$$L \leq 2\sqrt{3}|\mathbf{a} - \mathbf{b}|,$$

and pointed out that "$2\sqrt{3}$" could certainly be improved to about 3.33. They conjecture that

$$L \leq \tfrac{2}{3}\pi|\mathbf{a} - \mathbf{b}|.$$

This is true if C is any convex arc, with equality if C consists of two sides of a Reuleaux triangle, but has not been established for general curves.

The problem has obvious extensions to $\mathbb{R}^d$, with $d \geq 3$. Is it true that

$$L \leq d[2d/(d - 1)]^{1/2} \sin^{-1} ([(d - 1)/2d]^{1/2})|\mathbf{a} - \mathbf{b}|$$

with equality for the arc consisting of d consecutive sides of a Reuleaux simplex?

K. G. Binmore, On Turan's lemma, *Bull. London Math. Soc.* **3** (1971) 313–317; *MR* **44** #5434.

D. G. Larman & P. McMullen, Arcs with increasing chords, *Proc. Cambridge Philos. Soc.* **72** (1972) 205–207; *MR* **45** #4279.

G4. **Maximal sets avoiding certain distance configurations.** The well-known isodiametric theorem asserts that the largest plane set with no two points further than distance 1 apart, is the disk of unit diameter. J. Hewett asked for the maximum area of the plane set E such that whenever x_1, x_2, and x_3 are points of E, then at least one of the three distances $|x_1 - x_2|$, $|x_2 - x_3|$, and $|x_3 - x_1|$ is less than or equal to 1. Croft & Falconer showed that the largest set with this property consists of two disjoint (or touching) disks each of radius 1. What are the sets of largest volume with this property in d dimensions, with $d \geq 3$? Is the extremal a ball of radius $1/\sqrt{3}$ if d is large enough? If $d \geq 5$ this ball has larger d-dimensional content than two disjoint balls of radius $\tfrac{1}{2}$. What are the largest sets (for $d \geq 2$) if we require that whenever $x_1, \ldots, x_n$ are in E, $|x_i - x_j| \leq 1$ for at least m of the $\binom{n}{2}$ pairs? Croft &

Falconer point out that this is always equivalent to finding the largest set such that whenever $\mathbf{x}_1, \ldots, \mathbf{x}_k$ are in E, then $|\mathbf{x}_i - \mathbf{x}_j| \leq 1$ for *some* pair $\mathbf{x}_i$, $\mathbf{x}_j$, for a certain $k \leq n$. A compactness argument shows that extremal sets always exist, but it seems difficult to identify them.

The problem also has interesting features when E is required to be convex. Any set that can be covered by a pair of sets of diameter 1 clearly has the property that one of the distances realized by any three points is at most 1. The convex set of greatest area that can be covered by two sets of diameter 1 is given by the intersection of a parallel-sided strip of width $1/\sqrt{5}$ with two disks of radius 1 situated symmetrically on opposite edges of the strip. It is conjectured that this set is the convex set of greatest area satisfying the conditions of the original problem.

For many further problems associated with this circle of ideas, see the article by Croft & Falconer.

H. T. Croft & K. J. Falconer, On maximal Euclidean sets avoiding certain distance configurations, *Math. Proc. Cambridge Philos. Soc.* **89** (1981) 79–88; *MR* **81m**:52023.

G5.   Moving furniture around. Leo Moser asked for the region of largest area that can be moved around a right-angled corner in a corridor of width 1. (Think of this as a 2-dimensional problem.) The "Shephard piano" [see Figure G2(a)] with area 2.2074... is a good first approximation to the optimum. C. Francis modified this to get 2.21528... and iterating his refinement gives 2.21563.... Richard Guy obtained 2.21503... and combining his ideas with those of Francis gives 2.215649.... How close is this to the maximum area? Mike Guy and John Conway have pointed out that an extremal exists, but it is not known whether it is smooth or symmetric, or even unique.

Similarly, what is the largest region that can reverse in a T junction, with all roads of unit width? Here the "Conway car" [see Figure G2(b)] is a good initial approximation.

Maruyama considers further variations on this theme, and gives a general algorithm to find approximate solutions.

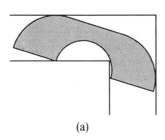
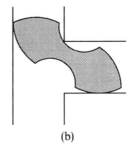

(a)   (b)

Figure G2.   Large pieces of furniture that can be moved round awkward corners. (a) The Shephard piano and (b) The Conway car.

What are the solutions to these problems if the region is required to be convex? Similarly, if C is a piece of wire with endpoints $\mathbf{x}$ and $\mathbf{y}$ that can be moved around the corner, how large can $|\mathbf{x} - \mathbf{y}|$ be? What about the higher dimensional analogs?

A related problem is discussed by Strang. It is possible to "post" a plane convex set of minimal width w through a slit of length l, if and only if $w \le l$. In three dimensions, however, Stark has constructed a convex "chair" K that can pass through a circular "door" D in a plane wall, but only by using a suitable sequence of rotations. Thus, no plane projection of K is contained in the disk D. Strang modifies this construction to get such an example in the case of a square door. Is it possible to find a convex body K that can be moved through a convex door D but with the area of every plane projection of K bigger than the area of D? If so, how large can the ratio (area of smallest plane projection of K)/(area of D) be?

What are the analogs of this problem for non-convex sets? Is there an algorithm for determining the smallest length of slit through which a plane set E may pass? Strang conjectures that if E has a smooth perimeter, then the chords of E that are "double normals" are the determining factors. One 3-dimensional version of this is the "prisoner's dilemma" of determining whether a (non-convex!) body will pass between equally spaced parallel bars.

A paper by Dawson introduces some general techniques applicable to mobility problems. The problem of finding an algorithm to determine how to move an object through a collection of obstacles has attracted much attention recently. For details and many further references, see, for example, Kedem & Sharir and Schwartz & Sharir.

R. J. McG. Dawson, On the mobility of bodies in R^n, *Math. Proc. Cambridge Philos. Soc.* **98** (1985) 403–412, corrigenda **99** (1986) 377–379; *MR* **87h**:52016a,b.

H. G. Debrunner & P. Mani-Levitska, Can you cover your shadows? *Discrete Comput. Geom.* **1** (1986) 45–58; *MR* **87e**:52010.

M. Goldberg, Solution to problem 66-11, *SIAM J.* **11** (1969) 76–78.

R. K. Guy, Monthly research problems, 1969–77, *Amer. Math. Monthly* **84** (1977) 807–815.

K. Kedem & M. Sharir, An efficient motion-planning algorithm for a convex polygonal object, *Discrete Comput. Geom.* **5** (1990) 43–75.

K. Maruyama, An approximation method for solving the sofa problem, *Inter. J. Comp. Inf. Sci.* **2** (1973) 29–48.

L. Moser, Problem 66-11, *SIAM Rev.* **8** (1966) 381.

J. T. Schwartz & M. Sharir, On the "piano mover's" problem, I, *Commun. Pure Appl. Math.* **36** (1983) 345–398; *MR* **84h**:70005; II, *Adv. Appl. Math.* **4** (1983) 298–351; *MR* **85h**:52014; III, *Int. J. Robotics Res.* **2** (1983) 46–75; *MR* **86a**:52016; V, *Commun. Pure Appl. Math.* **37** (1984) 815–848; *MR* **86j**:52010.

M. Sharir & E. Ariel-Sheffi, On the piano mover's problem, IV, *Commun. Pure Appl. Math.* **37** (1984) 479–493; *MR* **86j**:52009.

G. Strang, The width of a chair, *Amer. Math. Monthly* **89** (1982) 529–534; *MR* **84e**:52005.

N. R. Wagner, The sofa problem, *Amer. Math. Monthly* **83** (1976) 188–189; *MR* **53** #1422.

G6.   **Questions related to the Kakeya problem.** Besicovitch's construction of a plane set of area zero containing a unit line segment in every direction, was adapted to provide a solution to the Kakeya problem: that a unit segment can be rotated through an angle π by a continuous motion inside an arbitrarily small area. The ultimate solution seems to be Cunningham's remarkable result that this can be done inside a simply-connected subset of the unit disk with arbitrarily small area. For a full discussion of this problem and its ramifications see Falconer, Chapter 7 or Meschkowski, Chapter 8. We mention a few outstanding problems.

Wirsing asked us whether it is possible to move a circular arc of radius one and angle θ to a parallel position, keeping within a set of arbitraily small area. This was disproved by Davies. What, then, is the least area of the set in which such an arc can be moved to a parallel position a distance h away in the direction of the determining chord? Does a direct translation give the least area?

How large need the measure of a set be in order to be able to rotate a "fat needle," that is, a rectangle of sides one and ε? Presumably it tends to zero with ε, perhaps it is asymptotic to $c|\log \varepsilon|^{-1}$ as $\varepsilon \to 0$.

Let E be a bounded plane set of zero area that contains a line segment of length one in every direction. Given $\varepsilon > 0$, is it possible to find a sequence of strips (bounded by parallel lines) of widths $w_1, w_2, \ldots$ that cover E such that $\sum_{i=1}^{\infty} w_i < \varepsilon$? If this is not possible for all such E, is there at least some such set E for which such coverings exist? In asking this question, Rogers points out that if $s < 1$, then $\sum_{i=1}^{\infty} w_i^s \to \infty$ as $\sup_{1 \le i < \infty} w_i \to 0$.

The idea of a Perron tree (see Figure G3) is central to the construction of a Kakeya set. Take an equilateral triangle of height l, divide its base into k equal intervals, and join each of these to the opposite vertex to obtain k thin triangles. Now slide these along the base-line of the original triangles, so that they overlap as much as possible, that is, so that the total area A_k of the figure is a minimum. Schoenberg conjectures that A_k decreases monotonically as k increases, and if $k = 2^r$ then $A_k = 2/(r + 2)$, a value that is certainly attained for one configuration. He proves that $A_k = O(1/\log k)$; in fact all that one needs to solve the Kakeya problem is that $\lim \inf A_k = 0$ as $k \to \infty$. (Observe that each Perron tree contains a unit segment in the direction of every line joining a point of the base of the original triangle to the opposite vertex.) Progress on this problem would at least give an upper bound for the "fat needle" problem above.

The 3-dimensional situation is very different from that of the plane. Falconer and Marstrand have shown that a set that has a section of positive area parallel to every plane must have positive volume. What is the least volume of a set that contains some unit disk parallel to every plane? What is the least volume of a set that has a section of area of at least 1 parallel to every plane?

Another variation is to ask for the plane (measurable) set of least area that contains a congruent copy of every curve of length 1. This is the "non-

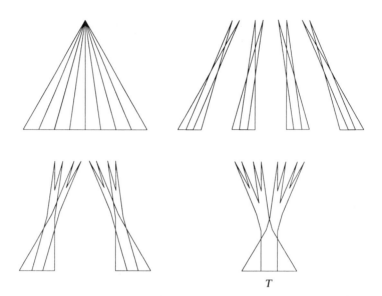

Figure G3.   Construction of a Perron tree T by sliding eight thin triangles along their common baseline.

convex" analog of the worm problem (see Section D18). A deep result of Marstrand implies that any such set must have positive area. What is the smallest such set if we insist that the curve is continuously deformable between any two of its positions within the set?

A very general, and presumably very difficult, problem would be to characterize those plane sets E for which there exists a set of zero area containing a congruent copy of E in every orientation. The Kakeya construction implies that a straight line is such a set, and the work of Davies shows that any countable union of straight lines has this property. Can general necessary or sufficient conditions be found?

A. S. Besicovitch, On Kakeya's problem and a similar one, *Math. Zeit.* **27** (1927) 312–320.

F. Cunningham, The Kakeya problem for simply connected and for star-shaped sets, *Amer. Math. Monthly* **78** (1971) 114–129; *MR* **43** #1044.

R. O. Davies, Some remarks on the Kakeya problem, *Proc. Cambridge Philos Soc.* **69** (1971) 417–421; *MR* **42** #7069.

K. J. Falconer, Continuity properties of k-plane integrals and Besicovitch sets, *Math. Proc. Cambridge Philos. Soc.* **87** (1980) 221–226; *MR* **81c**:53067.

K. J. Falconer, *The Geometry of Fractal Sets*, Cambridge University Press, Cambridge, 1985; *MR* **88d**:28001.

H. Meschkowski, [Mes].

J. M. Marstrand, Packing planes in R^3, *Mathematika* **26** (1979) 180–183; *MR* **81d**:52009.

J. M. Marstrand, Packing smooth curves in R^q, *Mathematika* **26** (1979) 1–12; *MR* **81d**:52009.

O. Perron, Über Satz von Besicovitch, *Math. Zeit.* **28** (1928) 383–386.

C. A. Rogers, Dimension prints, *Mathematika* **35** (1988) 1–27.

I. J. Schoenberg, On the Besicovitch–Perron solution of the Kakeya problem, *Studies in Mathematical Analysis*, Stanford University Press, Stanford, 1962, 359–363; *MR* **26** #4218.

I. J. Schoenberg, On certain minima related to the Besicovitch–Kakeya problem, *Mathematica (Cluj)* **4** (27) (1962) 145–148; *MR* **27** #4143.

G7. Measurable sets and lines.

Let E be a plane measurable set of positive but finite area. Croft asks whether there always exists a $\delta > 0$ such that for every t with $0 < t < \delta$, there is a line L with the linear measure of $L \cap E$ equal to t. Can we even do this in such a way that L varies continuously with t?

The 3-dimensional analog of this, with planes replacing lines, is certainly true. This follows from a result of Falconer, that the area of $P(\theta, t) \cap E$ is continuous in t for almost all θ, where $P(\theta, t)$ is the plane perpendicular to the unit vector θ and a distance t from the origin. Such results fail in two dimensions, as the Kakeya set [of area zero but containing a line segment in every direction (see Section G6)] shows, so a different approach will be required in the 2-dimensional case.

K. J. Falconer, Continuity properties of k-plane integrals and Besicovitch sets *Math. Proc. Cambridge Philos. Soc.* **87** (1980) 221–226; *MR* **81c**:53067.

H. T. Croft, Three lattice point problems of Steinhaus, *Quart. J. Math. Oxford Ser.* (2) **33** (1982) 71–83; *MR* **85g**:11051.

G8. Determining curves from intersections with lines.

Let C_1 and C_2 be rectifiable curves in the plane, and let $n_1(\theta, t)$ and $n_2(\theta, t)$ be the number of intersections of the curves with the line making an angle θ with a fixed axis and a distance t from the origin. Suppose that $n_1(\theta, t) = n_2(\theta, t)$ for all θ and t. Does this imply that C_1 and C_2 are identical? Horwitz points out that this is certainly so if C_1 and C_2 are twice differentiable curves.

The most natural setting for the problem is for **rectifiable sets**, that is subsets of a countable union of rectifiable arcs of finite total length (see for example Falconer, Chapter 3). Is such a set C uniquely determined to within a set of length zero if $n(\theta, t)$ is known for almost all (θ, t)? Further, how can C be reconstructed from this information?

Steinhaus defines a "distance" between curves by

$$d(C_1, C_2) = \frac{1}{2} \int_0^{2\pi} \int_0^\infty |n_1(\theta, t) - n_2(\theta, t)| \, dt \, d\theta$$

and examines some convergence properties of curves in this "metric," but it is unclear that $C_1 = C_2$ if $d(C_1, C_2) = 0$. Of course the length of C is given by $\frac{1}{2} \int_0^{2\pi} \int_0^\infty n(\theta, t) \, dt \, d\theta$ by Crofton's formula of integral geometry.

This question is a special case of a more general "tomography" problem. Let f be a bounded function supported by a rectifiable set C. Suppose that for each line L the sum of the values of f over the (almost always finite) set of points $C \cap L$ is given. Can f be determined from this information?

K. J. Falconer, *The Geometry of Fractal Sets*, Cambridge University Press, Cambridge, 1985; *M R* **88d**:28001.

A. Horwitz, Reconstructing, a function from its tangent lines, *Amer. Math. Monthly* **96** (1989) 807–813.

L. A. Santaló, *Integral Geometry and Geometric Probability*, Addison–Wesley, Reading, Mass., 1976; *M R* **55** #6340.

H. Steinhaus, Length, shape, and area, *Colloq. Math.* **3** (1954) 1–13; *M R* **16**, 121.

G9. Two sets which always intersect in a point.

Steinhaus asked whether there exist two subsets of the plane such that no matter how they are placed relative to each other, they intersect in a single point. (Of course, we except the trivial case of a single point and the whole plane!) Sierpiński used transfinite induction to show that such sets exist, assuming the axiom of choice. However, it would be interesting to have an explicit construction; in particular, can both sets be Borel sets? One can also look for pairs of sets that always intersect in exactly n points for $n > 1$. If n is even, a nice example is given by the collection of parallel lines with unit spacing, and the circle of radius $\frac{1}{4}n$ (see Figure G4); are there such simple examples if n is odd? The one-dimensional analog is easy: take the integers as one set and the interval $[0, n)$ as the other.

In a similar vein, Mazurkiewicz used the axiom of choice to construct a plane set that intersects every line in precisely two points. Sierpiński asked whether an effective construction is possible. Larman and Baston & Bostock showed that any F_σ set that cuts every line in at least one point, cuts some line in at least three points, so such sets cannot be F_σ. (Recall that an F_σ set is a countable union of closed sets.) Does a Borel (or even an analytic) set with this property exist?

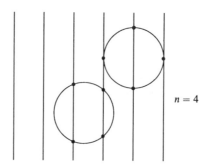

Figure G4.   A circle of unit radius always intersects a collection of parallel lines unit distance apart in exactly four points, however it is placed.

V. J. Baston & F. A. Bostock, On a theorem of Larman, *J. London Math. Soc.* (2) **5** (1972) 715–718; *M R* **47** #4138.

D. G. Larman, A problem of incidence, *J. London Math. Soc.* **43** (1968) 407–409; *M R* **38** #52.

S. Mazurkiewicz, *C. R. Sci. Lett. Varsovie* **7** (1914) 322–383.

W. Sierpiński, Cardinal and ordinal numbers, Warsaw, 1958, 446–447.

W. Sierpiński, Sur un problème de H. Steinhaus concernant les ensembles de points sur le plane, *Fund. Math.* **46** (1959) 191–194; *MR* **21** #851.

G10.   The chromatic number of the plane and of space.   A fascinating problem, apparently due to E. Nelson, that combines ideas from set theory, combinatorics, measure theory, and distance geometry, is to find the "chromatic number of the plane." More precisely, what is the smallest number of sets ("colors") with which we can cover the plane in such a way that no two points of the same set are a unit distance apart? In graph theoretic parlance, what is the chromatic number of the graph with vertices as the points of $\mathbb{R}^2$, with two vertices joined by an edge whenever they are at distance 1? This problem is purely finite in character, by the powerful graph-theoretic result of de Bruijn & Erdős: if we assume the axiom of choice, the chromatic number of a graph is the maximum chromatic number of its finite subgraphs.

Figure G5(a) shows a coloring of the plane with seven colors with no two points of the same color at distance 1 apart (the hexagons are regular with diameters slightly less than 1). Thus the **chromatic number** χ, that is, the minimum number of colors needed, is at most seven. On the other hand, $\chi \geq 4$: the reader may care to verify that the vertices of the configuration in Figure G5(b) cannot be colored with three colors without both ends of some edge having the same color (the edges all have length 1).

Surprisingly, these simple examples provide the best bounds known for χ in the general case. However, by placing fairly weak restrictions on the type of sets used to cover the plane, we can get better estimates. We denote the smallest number of sets which can cover the plane without any of the sets

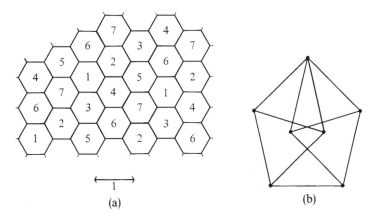

(a)   (b)

Figure G5.   (a) A coloring of the plane with seven colors with no pair of points distance one apart of the same color. (b) The "Moser spindle"—a configuration of seven points that cannot be colored using three colors without some pair of points at unit distance having the same color. (All the lines shown are of length 1.)

realizing unit distance by:

χ —if there is no restriction on the sets

χ_M—if the sets are (plane-Lebesgue) measurable

χ_C —if the sets are closed

χ_R —if the sets are unions of "regions," that is, bounded by rectifiable Jordan curves with finitely many regions intersecting any bounded portion of the plane, and with the boundary points considered to belong to some neighboring region (see Woodall for a more precise definition).

At the same time it is convenient to consider the more general problem of covering the plane with sets each not realizing *some* distance between its points (with possibly a different distance for each set). For each type of set listed above, we let ψ, ψ_M, ψ_C, and ψ_R denote the minimum possible number of sets. Thus, for example, if $\mathbb{R}^2$ is covered by fewer than ψ_M measurable sets, one of the sets realizes all distances. The following is known:

$$\psi \leq \chi, \quad \psi_M \leq \chi_M, \text{etc.,} \qquad \text{(trivial)}$$

$$\chi \leq \chi_R, \qquad \text{(trivial)}$$

$$4 \leq \chi \leq 7, \qquad \text{(examples above)}$$

$$4 \leq \psi, \qquad \text{(Woodall, Raiskii)}$$

$$\psi, \psi_M, \psi_C, \psi_R \leq 6, \qquad \text{(Woodall)}$$

$$6 \leq \chi_R \leq \chi_C, \quad \psi \leq \psi_R \leq \psi_C, \quad \text{(Woodall)}$$

$$\chi \leq \chi_M \leq \chi_C, \quad \psi \leq \psi_M \leq \psi_C, \quad \text{(easy)}$$

$$5 \leq \chi_M \qquad \text{(Falconer)}.$$

Thus we are left to resolve the following possibilities:

$$\chi = 4, 5, 6, \text{ or } 7,$$
$$\chi_M = 5, 6, \text{ or } 7,$$
$$\chi_C, \chi_R = 6 \text{ or } 7.$$
$$\psi, \psi_M, \psi_C, \psi_R = 4, 5, \text{ or } 6$$

The following problem for a single set is closely related. Define the density of a measurable subset E of the plane by $\lim_{r \to \infty} \text{area}(E \cap B_r)/\pi r^2$, where B_r is the ball of radius r, center the origin. Moser asked for the maximum density ρ of a set E that contains no pair of points unit distance apart. Clearly, $\chi_M \geq 1/\rho$; Croft constructed an example to show $\rho \geq 0.229\ldots$ by placing the intersection of a suitable disk and regular hexagon at each point of the equilateral triangular lattice. On the other hand, by translating Figure G5(b) across the plane, it follows that $\rho \leq 2/7 = 0.286\ldots$. In the same vein, Larman & Rogers ask what is the largest (measurable) subset of the disk of unit radius that contains no pair of points a unit distance apart? Surely it must be a disk of radius $\frac{1}{2}$, but this seems hard to prove. This would imply, by integrating over the plane, that $\rho \leq \frac{1}{4}$.

Of course, these questions may be asked in three or more dimensions. We denote by $\chi(\mathbb{R}^d)$, $\chi_M(\mathbb{R}^d)$, ... the minimum number of sets of a given type that are needed to cover $\mathbb{R}^d$ with no set realizing a distance 1, and by $\psi(\mathbb{R}^d)$, ... the minimum number if each set omits *some* distance. Questions of particular interest are the exact value of $\chi(\mathbb{R}^d)$ and $\chi_M(\mathbb{R}^d)$ for small values of d, and asymptotic estimates as $d \to \infty$.

An old result of Hadwiger is that $d + 1 < \psi_C(\mathbb{R}^d) \le \chi_C(\mathbb{R}^d)$; this was refined to $d + 1 < \psi(\mathbb{R}^d) \le \chi(\mathbb{R}^d)$ by Raiskii and Woodall. The first dramatic improvements in these estimates were due to Larman & Rogers, who pointed out that if there is a configuration of n points in $\mathbb{R}^d$ with the property that every $(k + 1)$-point subset of the configuration realizes unit distance, then $n/k \le \psi(\mathbb{R}^d) \le \chi(\mathbb{R}^d)$. By constructing suitable configurations they showed that $\frac{1}{6}n(n - 1) \le \psi(\mathbb{R}^d) \le \chi(\mathbb{R}^d)$. Frankl & Wilson devised an ingenious configuration that led to an exponential estimate $[1.2 + o(1)]^d \le \psi(\mathbb{R}^d) \le \chi(\mathbb{R}^d)$. The best known upper bound is due to Larman & Rogers who cover $\mathbb{R}^d$ by $[3 + o(1)]^d$ congruent closed sets, none of which realize unit distance. Székely has given a rather simpler decomposition into $c \cdot 11^d$ such sets. Thus

$$[1.2 + o(1)]^d \le \chi(\mathbb{R}^d) \le [3 + o(1)]^d$$

but better estimates would be of great interest. In particular does $\lim_{d \to \infty} \chi(\mathbb{R}^d)^{1/d}$ exist?

We list known estimates for small d; those not mentioned elsewhere are due to Székely.

$$4 \le \chi(\mathbb{R}^2) \le 7, \qquad 5 \le \chi_M(\mathbb{R}^2) \le 7,$$

$$5 \le \chi(\mathbb{R}^3) \le 21, \qquad 6 \le \chi_M(\mathbb{R}^3) \le 21,$$

$$6 \le \chi(\mathbb{R}^4), \qquad 8 \le \chi_M(\mathbb{R}^4).$$

A question that is still unsettled is whether $\chi(\mathbb{R}^d) = \chi_M(\mathbb{R}^d)$ for all d. Work by Székely suggests that equality may not be consistent with the axiom of choice.

The papers by Woodall and Székely & Wormold contain excellent surveys of these problems.

N. G. de Bruijn & P. Erdős, A color problem for infinite graphs and a problem in the theory of relations, *Indag. Math.* **13** (1951) 369–373; *MR* **13**, 763.

H. T. Croft, Incidence incidents, *Eureka (Cambridge)* **30** (1967) 22–26.

K. J. Falconer, The realization of distances in measurable subsets covering R^n, *J. Combin. Theory Ser. A* **31** (1981) 184–189; *MR* **82m**:05031.

P. Frankl & R. M. Wilson, Intersection theorems with geometric consequences, *Combinatorica* **1** (1981) 357–368; *MR* **84g**:05085.

H. Hadwiger, Ein Überdeckungssätz für den Euklidischen Raum, *Portugal. Math.* **4** (1944) 140–144.

H. Hadwiger, Ueberdeckung des Euklidischen Raumes durch kongruente Mengen, *Portugaliae Math.* **4** (1945) 238–242; *MR* **7**, 215.

H. Hadwiger, Ungelöste Problem Nr 40, *Elem. Math.* **16** (1961) 103–104.

H. Hadwiger, H. Debrunner & V. Klee, [HDK].

V. Klee, Some unsolved problems in plane geometry, *Math. Mag.* **52** (1979) 131–145; *MR* **80m**:52006.

D. G. Larman, A note on the realization of distances within sets in Euclidean space, *Comment. Math. Helv.* **53** (1978) 529–535; *MR* **80b**:52026.

D. G. Larman & C. A. Rogers, The realization of distances within Euclidean space, *Mathematika* **19** (1972) 1–24; *MR* **47** #7601.

L. Moser & W. Moser, Problem 10, *Canad. Math. Bull.* **1** (1958) 192, solution in *Canad. Math. Bull.* **4** (1961) 187–189.

W. Moser & J. Pach, Problem 61, [MP].

D. E. Raiskii, Realizations of all distances in a decomposition of the space R^n into $n + 1$ parts, *Mat. Zametki* **7** (1970) 319–323; *MR* **41** #6059.

L. A. Székely, Measurable chromatic number of geometric graphs and sets without some distances in Euclidean space, *Combinatorica* **4** (1984) 213–218; *MR* **85m**: 52008.

L. A. Székely, Remarks on the chromatic number of geometric graphs, in *Graphs and other Combinatorial Topics*, Teubner-Texte zur Mathematik 59, Leipzig, 1983, 312–315. *MR* **85e**:05076.

L. A. Székely & N. C. Wormold, Bounds on the measurable chromatic number of $\mathbb{R}^n$, *Discrete Math.* **75** (1989) 343–372.

D. R. Woodall, Distances realized by sets covering the plane, *J. Combin. Theory Ser. A* **14** (1973) 187–200; *MR* **46** #9868.

G11. Geometric graphs.

Geometric graphs generalize the ideas of the previous section. Let H be a (usually finite) set of distances and let G_H be the graph whose vertices are the points of the plane (more generally, of $\mathbb{R}^d$, $d \geq 1$) and with **x** "joined" to **y** if $|\mathbf{x} - \mathbf{y}|$ is a distance in H. We are interested in the graph theoretic properties of G_H, and in particular in the chromatic number $\chi(G_H)$. Thus $\chi(G_H)$ is the least number of colors required to color the plane so that no pair of points of the same color realize any distance in H. As before, the measurable chromatic number $\chi_M(G_H)$ pertains to the situation when the regions of each color must be measurable.

Erdős asks how large the chromatic number can be if the set H contains k distinct distances. Since the plane can be colored with seven colors that avoid any given distance, "intersection" of such colorings shows that $\chi(G_H) \leq \chi_M(G_H) \leq 7^k$. To get good estimates in the other direction one presumably would have to study (rather large) subgraphs of G_H. The square lattice provides an example with $\chi(G_H) \geq ck(\log k)^{1/2}$. Of particular interest is the rate at which $\max\{\chi(G_H) : |H| = k\}$ tends to infinity with k: is it exponential? Erdős suggested looking at this problem if the k distances in H all lie in a bounded interval, say $[1, 2]$. One would expect the maximum possible chromatic number to be much smaller in this case.

It would even be of interest to find *some* finite set of distances H for which $\chi(G_H)$ or $\chi_M(G_H)$ could be determined exactly. Székely has shown that $\chi_M(G_{\{1, \sqrt{3}\}}) \geq 6$. Is it true that $\chi(G_{\{1,t\}})$ is monotonic in t for $t > 1$, and equal to $\chi(G_{\{1\}})$, that is $\chi(\mathbb{R}^2)$, if t is close enough to 1? The same question can be asked for χ_M.

Let $\{r_i\}$ be a given decreasing sequence of distances covering to 0, and write $H(k) = \{r_1, r_2, \ldots, r_k\}$. Results of Falconer imply that $\chi_M(G_{H(k)}) \to \infty$ as $k \to \infty$. Can one obtain sensible estimates for the rate of convergence? Presumably $\chi(G_{H(k)}) \to \infty$ also.

Is it true that $\chi(G_H) = \chi_M(G_H)$ if H is finite? This is *not* true in the 1-dimensional case (coloring the real line) since $\chi_M(G_{\{1, \sqrt{2}\}}) = 3$, but, assuming the axiom of choice, $\chi(G_{\{1, \sqrt{2}\}}) = 2$. In fact, Székely exhibited a countable set H with $\chi(G_H) = 2$ but with $\chi_M(G_H)$ infinite. However, his examples are essentially 1 dimensional.

R. B. Eggleton, P. Erdős & D. K. Skilton, Colouring the real line, *J. Combin. Theory Ser. B* **39** (1985) 86–100; *MR* **87b**:05057.

P. Erdős, Combinatorial problems in geometry and number theory, in *Relations Between Combinatorics and other Parts of Mathematics*, Proc. Symp. Pure Math. 34, Amer. Math. Soc., 1970, 149–162.

P. Erdős & M. Simonovits, On the chromatic number of geometric graphs, *Ars Combin.* **9** (1980) 229–246; *MR* **82c**:05048.

K. J. Falconer, The realization of small distances in plane sets of positive measure, *Bull. London Math. Soc.* **18** (1986) 475–477; *MR* **87h**:28008.

L. A. Székely, Remarks on the chromatic number of geometric graphs, in *Graphs and other Combinatorial Topisc*, Teuber-Texte zur Mäthematik 59, Leipzig, 1983, 312–315; *MR* **85e**:05076.

L. A. Székely, Measurable chromatic number of geometric graphs and sets without some distance in Euclidean space. *Combinatorica* 4 (1984) 213–218; *MR* **85m**:52008.

G12. **Euclidean Ramsey problems.** Euclidean Ramsey theory refers to the geometrical analog of an important area of graph theory known as Ramsey theory. A finite subset S of $\mathbb{R}^d$ is called *n*-**Ramsey** if, however $\mathbb{R}^d$ is colored with n colors, there is a set S' that is congruent to S with all points of S' the same color. A set that is *n*-Ramsey for every positive integer n is called **Ramsey**.

These notions were introduced by Erdős et al., who have made considerable progress towards characterizing the sets S that are Ramsey or *n*-Ramsey. In particular, they have shown that if $S \subset \mathbb{R}^d$ is Ramsey, then the points of S must lie on a spherical surface. On the other hand, any subset of the vertices of a "brick" (i.e., a rectangular parallelepiped) is Ramsey. What is the exact characterization of Ramsey sets in the plane or in $\mathbb{R}^d$? Is it either of these alternatives, or is it somewhere in between? The vertices of any triangle is Ramsey in $\mathbb{R}^2$, but no complete characterization of Ramsey quadrilaterals is known, and no pentagon has yet been shown to be Ramsey.

Which triangles are 2-Ramsey (i.e., have congruent copies with either three red vertices or three blue vertices for every coloring of the plane with red and blue? Equilateral triangles are not 2-Ramsey (consider a coloring of the plane by alternate colored stripes of width equal to the height of the triangle). Is every non-equilateral triangle in the plane 2-Ramsey? Shader has shown that every right-angled triangle is.

There are many variations on this type of problem. Juhász showed that, if the plane is colored red and blue, then either there are two blue points at unit distance apart, or the red set contains a congruent copy of the vertices of every quadrilateral. Erdős, et al., showed that this is false if "quadrilateral" is replaced by "*n*-gon" for $n \geq 10^{12}$ and Juhász reduced this to $n \geq 12$. Is it true for $n = 5$ or, indeed any $n < 12$?

P. Erdős, R. L. Graham, P. Montgomery, B. L. Rothschild, J. H. Spencer & E. G. Straus, Euclidean Ramsey theorems I, II, III, *J. Combin. Theory Ser. A* **14** (1973) 341–363, *Colloq. Math. Soc. János Bolyai* **10** (1973) 529–557, 559–583; *MR* **47** #4825, **52** #2935, **52** #2936.

P. Frankl & V. Rödl, All triangles are Ramsey, *Trans. Amer. Math. Soc.* **297** (1986) 777–779; *MR* **88d**:05018.

P. Frankl & V. Rödl, Forbidden intersections, *Trans. Amer. Math. Soc.* **300** (1987) 259–286; *MR* **88m**:05003.

P. Frankl & V. Rödl, A partition property of simplices in Euclidean spaces, *J. Amer. Math. Soc.* **3** (1990) 1–7.

R. L. Graham, On partitions of E^n, *J. Combin. Theory Ser. A* **28** (1980) 89–97; *MR* **81g**:05021.

R. L. Graham, Old and new Euclidean Ramsey theorems, [GLMP], 20–30; *MR* **87b**:05021.

R. L. Graham, B. L. Rothschild & J. H. Spencer, *Ramsey Theory*, Wiley, New York, 1980; *MR* **82b**:05001.

R. Juhász, Ramsey-type theorems in the plane, *J. Combin. Theory Ser. A* **27** (1979) 152–160; *MR* **81f**:05125.

L. Shader, All right triangles are Ramsey in $\mathbb{E}^2$! *J. Combin. Theory Ser. A* **20** (1976) 385–389.

G13. Triangles with vertices in sets of a given area.

The following is an old and difficult problem of Erdős: is there an absolute constant c (ultimately as small as possible, perhaps even $4\pi/3\sqrt{3}$, the area of the circumcircle of an equilateral triangle of unit area) such that any (measurable) plane set E of area c contains the vertices of a triangle of area 1?

If E has infinite area, or even has positive area and is unbounded, then it has this property. To see this, choose perpendicular x and y axes that each cut E in positive length. By the well-known result of Steinhaus, that sets of positive measure realize all sufficiently small distances, we have that $E \cap \{x \text{ axis}\}$ and $E \cap \{y \text{ axis}\}$ contain all distances in $(0, \delta_1)$ and $(0, \delta_2)$, respectively. Thus any point of E outside the rectangle $\{(x, y) : |x| \leq 2/\delta_1, |y| \leq 2/\delta_2\}$ together with two points on the appropriate axis, give a triangle of area 1.

Suppose now that E has infinite area. By the preceding remarks we can find in E the vertices of a triangle of area 1, but can we demand that the triangle be right-angled or, alternatively, isosceles? If E has infinite area, Erdős points out that one can always find four points of E forming a trapezoid (a quadrilateral with one pair of sides parallel) of unit area, but can one find vertices of a convex polygon of area 1 and with all sides equal? There are clearly many variations on this theme.

P. Erdős, Some combinatorial, geometric and set theoretic problems in measure theory, in *Measure Theory*, Oberwolfach, 1983, 321–327; Lecture Notes in Math. 108, Springer, Berlin, 1984.

G14. Sets containing large triangles.

The following elegant result was conjectured by Székely: Let $E \subset \mathbb{R}^2$ be a (measurable) set with "positive density at infinity," that is, with

$$\limsup_{r \to \infty} \operatorname{area}(E \cap B_r)/r^2 > 0.$$

Then there exists a number d_0 such that for any $d \geq d_0$ we may find $\mathbf{x}, \mathbf{y} \in E$ with $|\mathbf{x} - \mathbf{y}| = d$. Proofs of this were given by Falconer & Marstrand, and also by Bourgain.

However, the following natural generalization is still open. Do all plane sets E with positive density at infinity contain a congruent copy of the vertices of every sufficiently large equilateral triangle? More generally, given three noncollinear points $\{\mathbf{x}_1, \mathbf{x}_2, \mathbf{x}_3\}$, does such an E contain a set congruent to $\{r\mathbf{x}_1, r\mathbf{x}_2, r\mathbf{x}_3\}$ for every sufficiently large r? Examples show that this is false if the points are collinear, and false for 4-point sets.

Now let $f(r)$ be the smallest number such that every (measurable) set E of area greater than $f(r)$ in the disk of radius r contains the vertices of an equilateral triangle of side greater than 1. What can be said about $f(r)$ when r is large? Erdős conjectured that $f(r)/r^2 \to 0$ as $r \to \infty$, and Fürstenberg pointed out that this follows from a generalization of Szemerédi's theorem on sequences containing arbitrarily long arithmetic progressions. On the other hand, it is possible to adopt a method of Falconer used in a problem involving the combinatorial geometry of fractals to show that $f(r) \geq cr^{2-\varepsilon}$ for some constant c for any $\varepsilon > 0$. What is the exact asymptotic form of f?

J. Bourgain, A Szemerédi type theorem for sets of positive density in R^k, *Israel J. Math.* **54** (1986) 307–316; *MR* **87j**:11012.

P. Erdős, Some combinatorial, geometric and set theory problems in measure theory, *Measure Theory*, Oberwolfach 1983, Lecture Notes in Math. 1089, Springer-Verlag, 321–327.

K. J. Falconer, A problem of Erdős on fractal combinatorial geometry, *J. Combin. Theory. Ser. A.*, to appear.

K. J. Falconer & J. M. Marstrand, Sets with positive density at infinity contain all large distances, *Bull. London Math. Soc.* **18** (1986) 471–474; *MR* **87h**:28007.

G15. Similar copies of sequences. If $S \equiv \{x_1, x_2, \ldots, x_k\}$ is a finite set of real numbers, then it is not hard to see that every measurable subset of $\mathbb{R}$ of positive Lebesgue measure contains a subset geometrically similar to S. This is a generalization of the result of Steinhaus, that, given a set E of positive measure, there is a number $\delta > 0$ such that if $0 < a < \delta$ then there is a pair of points in E a distance a apart.

Now let $S \equiv \{x_1, x_2, \ldots\}$ where the infinite sequence of positive numbers $\{x_i\}$ decreases to 0. An old "\$100 problem" of Erdős is to show that there is always a measurable subset E of $\mathbb{R}$ of positive Lebesgue measure which contains no similar copy of S.

Various authors have proved this if $\{x_i\}$ converges sufficiently slowly, for example, with $\limsup_{i \to \infty} \{x_{i+1}/x_i\} = 1$, and Bourgain has dealt with the case where $S = S_1 + S_2 + S_3$ (i.e., $S = \{s_1 + s_2 + s_3 : s_i \in S_i\}$) where S_1, S_2, and S_3 are infinite sets. However, the general result remains unproved; it is not even known if it is true for every uncountable set S.

Some variants of this problem concern Baire category rather than measure, (e.g., see Arias de Reyna).

J. Arias de Reyna, Some results connected with a problem of Erdős III, *Proc. Amer. Math. Soc.* **89** (1983) 291–292; *MR* **85d**:28001.

J. Bourgain, Construction of sets of positive measure not containing an affine image of a given infinite structure, *Israel J. Math.* **60** (1987) 333–344; *MR* **89g**:28004.

S. J. Eigen, Putting convergent sequences into measurable sets, *Studia Sci. Math. Hungar.* **20** (1985) 411–412; *MR* **88f**:28003.

P. Erdős, My Scottish Book "problems," in [Mau], 35–43.

K. J. Falconer, On a problem of Erdős on sequences and measurable sets, *Proc. Amer. Math. Soc.* **90** (1984) 77–78; *MR* **85e**:28008.

P. Komjáth, Large sets not containing images of a given sequence, *Canad. Math. Bull.* **26** (1983) 41–43; *MR* **85d**:28003.

G16.   Unions of similar copies of sets.

Let E be a measurable set of positive real numbers of positive length (Lebesgue measure) and let $F = \bigcup_{r=1}^{\infty} rE$ (where rE is the set $\{rx : x \in E\}$). (Thus x is in F if x/r is in E for some integer r). Haight asks if, for almost all positive x (i.e., for all x except in a set of length zero) we can find a number $n(x)$ such that if $n \geq n(x)$, then $nx \in F$.

Note that this is false if, instead, we take $F = \bigcup_{r=1}^{\infty} (r + E)$; as shown by Marstrand's counterexample to Khinchin's conjecture.

J. M. Marstrand, On Khinchin's conjecture about strong uniform distribution, *Proc. London Math. Soc.* (3) **21** (1970) 540–556; *MR* **45** #185.

J. Haight, A linear set of infinite measure with no two points having integral ratio, *Mathematika* **17** (1970) 133–138; *MR* **42** #3046.

Index of Authors Cited

The names appearing here are those of authors whose works are referred to in this volume. Standard References, on page xv, are denoted by SR, those listed under Other Problem Collections, page xiii, by P, and those at the end of the section on Notations and Definitions, page 1, by N. References in the Chapter introductions are denoted by A, B etc., and in the problem sections by the section number (e.g., C11). Mentions of names unsupported by references are listed in the General Index.

Abreth, O.   B9
Adler, I.   B20
Adhikari, A.   A28
Affentranger, F.   B5
Ajtai, M.   D6
Aleksandrov, A.D.   B, B21
Alexander, R.   B13, D13, E6, F17
Alikaski, H.A.   B5
Alon, N.   B15, F11
Altshuler, A.   B15
Ament, P.   D1
Amilibia, A.M.   A20
Appel, K.   B19
Arias de Reyna, J.   G15
Ariel–Sheffi, E.   G5
Arkinstall, J.R.   E13
Armstrong, W.W.   A35
Asimov, L.   B14
Asplund, E.   A16
Auerbach, H.   A6

Baddeley, A.   B5
Bagemihl, F.   E7
Baker, A.   F
Baker, B.S.   C7
Ball, K.   A9, D13

Bambah, R.P.   E14
Banach, S.   C20
Bandle, C.   A17
Bang, T.   D13
Bannai, E.   F3
Bárány, I.   B5, E1, F8, F11, F13
Barnes, F.W.   C16
Barnette, D.   B16, B20, B24
Barnsley, M.   C17
Baston, V.J.D.   E7, G9
Beck, J.   E11, F1, F2, F12, F17
Behrend, F.   A16
Bender, E.A.   B15, C16, E12
Benson, R.V.   A
Berger, R.   C14, C16
Berlekamp, E.R.   B25
Berman, J.D.   F17
Berry, M.V.   A4
Besicovitch, A.S.   A15, A23, D18, E7, G6
Bessaga, C.   A36
Betke, U.   D9
Bezdek, A.   A20, B1
Bezdek, K.   A20, B1, C13, D1, D3, D18, E12
Bieri, H.   A19, A21, B6
Billera, L.J.   B17

185

General Index

Names appear here if their mention in this volume is unsupported by references. Single letter entries refer to the introductory section Notation and Definitions, N (pages 1 to 5), and to the introductions to Chapters A to G.